STUDY GUIDE

Human Biology
Concepts and Current Issues

Third Edition

Judith Stewart
Community College of Southern Nevada

San Francisco Boston New York
Cape Town Hong Kong London Madrid Mexico City
Montreal Munich Paris Singapore Sydney Tokyo Toronto

Publisher: Daryl Fox
Senior Acquisitions Editor: Deirdre McGill
Editorial Assistant: Alison Rodal and Sabrina Larson
Managing Editor: Wendy Earl
Production Editor: Leslie Austin
Composition: Cecelia G. Morales
Manufacturing Buyer: Stacey Weinberger
Senior Marketing Manager: Lauren Harp
Cover Designer: Frances Baca

Copyright © 2006 by Pearson Education, Inc., publishing as Pearson Benjamin Cummings, San Francisco, CA 94111. All rights reserved. Printed in the United States of America. This publication is protected by Copyright and permission should be obtained from the publisher prior to any prohibited reproduction, storage in a retrieval system, or transmission in any form or by any means, electronic, mechanical, photocopying, recording, or likewise. For information regarding permission(s), write to: Pearson Education, Inc., Rights and Permissions Department.

Many of the designations used by manufacturers and sellers to distinguish their products are claimed as trademarks. Where those designations appear in this book, and the publisher was aware of a trademark claim, the designations have been printed in initial caps or all caps.

ISBN 0-8053-7199-0

1 2 3 4 5 6 7 8 9 10—TCS—08 07 06 05 04

www.aw-bc.com

Preface

"Argue for your limitations, and you shall have them."

As you begin your study of science, one of the most important skills to acquire is the belief that you can succeed. Combined with commitment, adequate study time, practice, and utilization of available resources, you can master the concepts presented in your textbook. As you approach the *Study Guide,* be sure you have thoroughly read and studied the text material. Select a quiet location. Work through the exercises first without referring to the textbook or your notes. Use section references that accompany the exercises only to clarify a question you cannot answer, or have answered incorrectly. The *Study Guide* will be of little value if used as a worksheet, filling in answers while referring to the text. Use the *Study Guide* to practice applying what you have already learned through studying the textbook, and the Chapter Test to predict your readiness for classroom exams.

Each chapter of the *Study Guide* includes the following components:

Chapter Summary and Objectives: A list of 20 key concepts presented in the textbook.

Exercises: A combination of matching, fill-in-the-blank, labeling, completion, short answer, word choice, and crossword puzzles that test your understanding of the topic. Answers to most exercises are provided in the answer key located at the end of the *Study Guide*.

Chapter Test: A selection of 25 multiple-choice questions that assess your mastery of the topic and predict your performance on an exam. Take the Chapter Test only after you have successfully completed the exercises.

Key Concept Review Questions: The 20 key concepts listed at the beginning of each *Study Guide* chapter are rewritten as questions, assessing your comprehension of the main topics presented in the text. Answers to these questions are found by referring back to the Chapter Summary and Objectives.

Schedule time to study on a regular basis, preferably daily. You may choose to study an entire chapter first prior to working the exercises, or you may study the textbook in sections that correspond to the sections in the *Study Guide*. Either way, be sure to review the material several times to achieve a thorough understanding. Learning science can change your life, your perspective on many societal issues, and your ability to make informed choices. It all begins with you.

JUDY STEWART

Contents

Chapter 1	Human Biology, Science, and Society	1
Chapter 2	The Chemistry of Living Things	14
Chapter 3	Structure and Function of Cells	28
Chapter 4	From Cells to Organ Systems	42
Chapter 5	The Skeletal System	58
Chapter 6	The Muscular System	71
Chapter 7	Blood	84
Chapter 8	Heart and Blood Vessels	97
Chapter 9	The Immune System and Mechanisms of Defense	113
Chapter 10	The Respiratory System: Exchange of Gases	127
Chapter 11	The Nervous System—Integration and Control	142
Chapter 12	Sensory Mechanisms	158
Chapter 13	The Endocrine System	170
Chapter 14	The Digestive System and Nutrition	183
Chapter 15	The Urinary System	199
Chapter 16	Reproductive Systems	214
Chapter 17	Cell Reproduction and Differentiation	229
Chapter 18	Cancer: Uncontrolled Cell Division and Differentiation	243
Chapter 19	Genetics and Inheritance	255
Chapter 20	DNA Technology and Genetic Engineering	268
Chapter 21	Development and Aging	280
Chapter 22	Evolution and the Origins of Life	294
Chapter 23	Ecosystems and Populations	307
Chapter 24	Human Impacts, Biodiversity, and Environmental Issues	320
	Answer Key	333

Human Biology, Science, and Society

Chapter Summary and Key Concepts

After reading and studying this chapter you should know the following:

Sections 1.1, 1.2

1. Biology is the study of life.
2. Living things can be characterized by their molecular composition, energy requirements, cellular nature, homeostatic mechanisms, responsiveness, ability to grow and reproduce, and evolution.
3. A system of classification categorizes living things according to their characteristics.
4. As living organisms, humans are characterized by bipedalism, opposable thumbs, a large brain, and complex language skills.
5. Levels of biological organization exist in living things, beginning with atoms and becoming progressively more complex and inclusive.

Sections 1.3, 1.4

6. Scientific knowledge is information about the natural world and is gained by practicing the scientific method.
7. The scientific method is a series of steps that allow scientists to gather information about the natural world and test their ideas.
8. The scientific method begins with an observation and is followed by a hypothesis, a prediction, an experiment, data, and a conclusion.
9. A well-supported hypothesis becomes a theory.
10. Scientists publish their work in peer-reviewed journals to allow other scientists to verify their findings.
11. Scientific information is presented in many forms; it is important to be aware of different degrees of accuracy in each type of media.

Sections 1.5, 1.6

12. Good scientists use the tools of critical thinking.
13. Informed individuals can become critical thinkers.
14. Science uses statistics to interpret data.
15. A graph is one effective way to present information.
16. Facts differ from anecdotes and conclusions because facts are verifiable.
17. Correlation is a pattern between two variables, and it does not necessarily indicate causality.
18. Technology is the application of science and has a great potential to improve our world.
19. Science is an explanation of observable events in a physical world; it cannot be used to explain or support feelings, opinions, or abstract ideas.
20. A knowledge of science helps individuals to make informed choices.

Exercises

Complete the exercises for each section after you have read and studied the section. If you cannot answer some questions, or answer them incorrectly, return to the chapter and review this information. You may find it helpful to work on only one section at a time. When you have completed all sections, take the Chapter Test as an indicator of your mastery of this topic.

1.1 The characteristics of life

1.2 How humans fit into the natural world

Matching

_____ 1. **science** a. one of two domains comprised of bacteria

_____ 2. **biology** b. the second smallest unit of classification

_____ 3. **metabolism** c. five main categories used to classify living things

_____ 4. **cell** d. the study of life

_____ 5. **homeostasis** e. organisms that can interbreed in nature and produce fertile offspring

_____ 6. **kingdoms** f. a process that allows plants to capture energy from sunlight

_____ 7. **Eukarya** g. a kingdom containing unicellular and simple multicellular eukaryotes

_____ 8. **photosynthesis** h. a level of classification that encompasses kingdoms

_____ 9. *Protista* i. physical and chemical processes involving the energy transformations that maintain life

_____ 10. **domain** j. the maintenance of a constant internal environment in living things

_____ 11. **Bacteria** k. in the three-domain system of classification, this domain comprises all organisms with cells containing a nucleus

_____ 12. **species** l. the smallest unit of life that exhibits all the characteristics of life

_____ 13. **genus** m. the study of the natural world

Fill-in-the-Blank

Referenced sections are indicated in parentheses.

14. Organisms possessing cells that lack a nucleus are called _____. (1.1)

15. Organisms possessing cells that have a nucleus are called _____. (1.1)

16. In the five-kingdom classification system, the five kingdoms are _____, _____, _____, _____, and _____. (1.2)

17. Animals with a nerve cord and a backbone are called _____. (1.2)

18. Vertebrates with mammary glands are classified as _____. (1.2)

19. Humans, apes, and monkeys are all considered mammals and are further classified as _____. (1.2)

20. The ability to stand upright and walk on two legs is called _____. (1.2)

21. Humans have _____ thumbs, allowing them to grasp small objects with precision. (1.2)

22. Humans have a large _____ mass in relation to their body size. (1.2)

23. The capacity for complex _____ is a defining feature of being human. (1.2)

Completion

In the table below, identify the characteristic of life that is described.

Example	Characteristic of Life
24. Living organisms are composed of carbohydrates, proteins, lipids, and nucleic acids.	
25. At one time, giraffes had short necks. Today all giraffes have long necks.	
26. Viruses are not cellular in composition, and they are not considered to be living.	
27. Plants use the energy of the sun to manufacture carbohydrates, while humans digest carbohydrates to obtain energy.	
28. Most individuals dress warmly in the winter.	
29. When you are cold you often develop "goose bumps" as a result of tiny muscles under the skin contracting to produce heat.	
30. Single-celled organisms use the process of cell division to create new cells.	

Ordering

Place the following levels of biological organization in the proper sequence, beginning with the simplest and ending with the most complex.

a. tissue d. organ system g. population
b. organism e. community h. atom
c. cell f. organ i. ecosystem

31. _____ 32. _____ 33. _____
34. _____ 35. _____ 36. _____
37. _____ 38. _____ 39. _____

Short Answer

40. When are the characteristics of life not applicable in describing living things?

41. How is a domain different from a kingdom?

42. List the four defining features of humans.

1.3 Science is both a body of knowledge and a process

1.4 Sources of scientific information vary in style and quality

1.5 Learning to be a critical thinker

1.6 The role of science in society

Matching

_____ 1. **scientific knowledge** a. a broad hypothesis that has been repeatedly tested and supported over time

_____ 2. **scientific method** b. a false treatment given to a control group in an experiment

_____ 3. **hypothesis** c. a factor that might vary during the course of an experiment

_____ 4. **experiment** d. the application of science

_____ 5. **variable** e. the group in a controlled experiment that is not exposed to the experimental treatment

_____ 6. **experimental group** f. a questioning attitude required to practice science

_____ 7. **control group** g. organized, reliable information about the natural world

_____ 8. **placebo** h. a tentative statement about the natural world

continued

_____ 9. **theory**   i. a carefully planned and executed test of a prediction

_____ 10. **skepticism**   j. the group in a controlled experiment that is exposed to the experimental treatment

_____ 11. **technology**   k. the process used to acquire scientific knowledge

Fill-in-the-Blank

Referenced sections are indicated in parentheses.

12. Scientific knowledge relies on _____ and _____. (1.3)

13. When we use observations to make broad generalizations, we are using _____ reasoning. (1.3)

14. A hypothesis is a statement that can lead to _____ deductions. (1.3)

15. A working hypothesis is a testable _____. (1.3)

16. In _____ reasoning, a generalization is applied to specific cases. (1.3)

17. The variable of interest in an experiment is called the _____ variable. (1.3)

18. When the subjects in an experiment do not know who is in the control group and who is in the experimental group, this is a _____ study. (1.3)

19. When neither the subjects nor the administrators of an experiment know who is in the control group and who is in the experimental group, this is a _____ study. (1.3)

20. When a prediction is found to be false, the _____ must be modified. (1.3)

21. A hypothesis cannot be _____, it can only be _____ or _____. (1.3)

22. The scientific method can be described as a process of _____. (1.3)

23. Scientists publish their findings in _____ journals so that other scientists can examine their work. (1.3)

24. Numerical information is called _____. (1.5)

25. The range above and below a numerical average that scientists believe represents accurate data is called the _____. (1.5)

26. A testimonial is an example of _____ evidence. (1.5)

27. Information that can be verified is a _____. (1.5)

28. A relationship between two variables is a _____, but it does not always mean that one variable _____ the other. (1.5)

29. Science is limited to the explanation of _____ events in the _____ world. (1.6)

Completion

Scientific information is reported through many types of media. For each type listed below, name one strength and one weakness of this resource for a general biology student. (There are no answers to this exercise in the answer key. Refer to Section 1.4 of your text to check your responses.)

Type of Media	Strengths	Weaknesses
Scientific journal		
Popular science magazine		
Newspaper		
Web site ending in .com		
Web site ending in .edu		

Labeling

In Figure 1.1, indicate which of the following terms apply to indicated areas on the graph.

a. ordinate
b. abscissa
c. independent variable
d. dependent variable
e. standard error bar

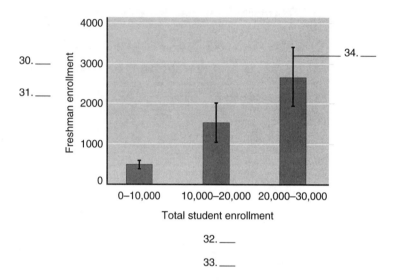

Figure 1.1

Chapter Test

Multiple Choice

1. Biology is the study of:
   a. science.
   b. ecosystems.
   c. life.
   d. biochemistry.

2. Which of the following are molecules used by all living things?
   a. proteins
   b. carbohydrates
   c. nucleic acids
   d. lipids
   e. all of the above

3. Living things require energy. The processes involved in acquiring and transforming energy are collectively referred to as:
   a. chemistry.
   b. physical biochemistry.
   c. metabolism.
   d. catabolism.

4. The absence of _____ always indicates that something is "nonliving."
   a. DNA
   b. evolution
   c. reproduction
   d. cells

5. Which of the following characteristics of life cannot be observed in a single organism, such as a human being?
   a. homeostasis
   b. growth and reproduction
   c. evolution
   d. responsiveness to the external environment

6. Classification of organisms in the five kingdom systems depends on all of the following *except:*
   a. the presence or absence of a nucleus.
   b. the type of metabolism.
   c. the number of cells.
   d. the type of reproduction.

7. Single-celled prokaryotic organisms are placed in the domain:
   a. Bacteria.
   b. Archaea.
   c. Eukarya.
   d. a and b.

8. The defining features of humans include all of the following *except:*
   a. opposable thumbs.
   b. bipedalism.
   c. sexual reproduction.
   d. large brain mass in relation to body size.

9. A group of tissues that carry out a specific function constitute a(n):
   a. cell.
   b. organism.
   c. organ.
   d. community.

10. Place the following steps of the scientific method in the correct sequence in which they would be used.
    I. formulate a hypothesis
    II. observe and generalize
    III. experiment
    IV. make a testable prediction
    V. modify hypothesis if necessary

    a. I, V, II, IV, III
    b. II, I, IV, III, V
    c. II, IV, III, I, V
    d. IV, I, II, III, V

11. In a scientific experiment, a placebo is given to the:
    a. experimental group.
    b. control group.
    c. both groups at different times.
    d. neither group.

12. You wish to test the effect of different amounts of a particular fertilizer on the growth of tomato plants. In your experiment, the controlled variable is:
    a. the type of fertilizer.
    b. plant growth.
    c. the amount of fertilizer.
    d. the type of tomato plant used.

13. The most accurate results will be obtained by a scientific study that:
    a. has a control group and an experimental group.
    b. is "double-blind."
    c. is "blind."
    d. identifies variables that may affect the outcome of the experiment.

14. Which of the following does not describe a prediction?
    a. an "if...then" statement
    b. a vague, general statement
    c. a testable statement
    d. all of the above describe a prediction

15. Confidence in a hypothesis is developed when:
    a. many scientists propose the hypothesis at the same time.
    b. the experiment designed to test the hypothesis has a control group and an experimental group.
    c. the results of an experiment support the hypothesis.
    d. many scientists have tried and failed to disprove the hypothesis.

16. The goal of science is to _____ what is _____.
    a. prove, truth
    b. prove, false
    c. disprove, truth
    d. disprove, false

17. Scientists most often publish their findings:
    a. in popular magazines.
    b. in peer-reviewed journals.
    c. in documentaries.
    d. on the Internet.

18. A fact is:
    a. a judgment.
    b. an interpretation.
    c. verifiable information.
    d. numerical data.

19. Statistics is the mathematics of organizing and interpreting:
    a. graphs.
    b. experiments.
    c. variables.
    d. data.

20. You begin eating broccoli every night of the week, and notice that you are more energetic. Which of the following would be the most accurate statement?
    a. Eating broccoli causes you to be more energetic.
    b. Eating broccoli is a good way to combat fatigue.
    c. Eating broccoli appears to be correlated with increased energy.
    d. Eating broccoli provides nutrients that are causing you to have more energy.

21. Developing the skills of a critical thinker will most help you to:
    a. understand information.
    b. question information.
    c. generate information.
    d. prove information.

22. Science cannot be used to explain:
    a. data.
    b. observations of the physical world.
    c. nonphysical events outside the natural world.
    d. science can be used to explain all of the above.

23. Anecdotal evidence often appears as:
    a. a double-blind study.
    b. a testimonial by a nonscientist.
    c. data.
    d. statistical evidence.

24. In a graph, the standard error is represented by:
    a. the ordinate.
    b. the abscissa.
    c. small lines that extend above and below the average number.
    d. bars or dots indicating the average of all data collected.

25. Understanding science and the scientific method can help individuals:
    a. make informed decisions.
    b. contribute to society.
    c. understand the impact of technology.
    d. evaluate the validity of nonphysical events that do not fall within the limits of the natural world.

Key Concept Review Questions

Each of the Key Concepts listed at the beginning of this chapter has been rewritten as a question below. After successively completing the study guide exercises and the Chapter Test, you should be able to answer each of these questions. Refer to the Key Concepts list at the beginning of this chapter to check your answers.

1. What is biology?
2. What are the seven characteristics of living things?
3. How do classification systems categorize living things?
4. What are the characteristics of humans?
5. What are the levels of biological organization?
6. What is scientific knowledge, and how is it gained?
7. What is the scientific method?
8. What are the steps of the scientific method?
9. What happens to a well-supported hypothesis?
10. Where do scientists most often publish their work?
11. When reading or watching the presentation of scientific information, what should you be aware of?
12. What type of thinking is used by good scientists?
13. Is this type of thinking limited to scientists?

14. What is the value of statistics to science?
15. What is a graph?
16. How do facts differ from anecdotes and conclusions?
17. What is a correlation? How does it relate to causality?
18. What is technology? What is the value of technology?
19. What are the limitations of science?
20. How is a knowledge of science valuable to individuals?

2

The Chemistry of Living Things

Chapter Summary and Key Concepts

After reading and studying this chapter you should know the following:

Sections 2.1, 2.2

1. Matter is composed of elements; elements are composed of atoms; atoms are composed of protons, neutrons, and electrons.

2. The atomic number represents the number of protons in an atom; the mass number represents the total number of protons and neutrons in an atom's nucleus.

3. Isotopes are atoms of an element that have more or less neutrons than typical atoms.

4. Energy is the ability to do work. Electrons have potential energy based on their position in an electron shell.

5. Chemical bonds form between atoms that need to fill their outermost occupied electron shell.

6. Covalent bonds form when atoms share electrons; ionic bonds form between oppositely charged ions; hydrogen bonds form between polar molecules.

Sections 2.3, 2.4

7. Water is an effective biological solvent because the water molecules are polar, and water is a liquid at body temperature.

8. Water helps to regulate body temperature because it can absorb and hold heat.

9. An acid is a molecule that donates H^+ to a solution. A base is a molecule that absorbs H^+ in a solution.

10. The pH scale is a measure of the hydrogen ion concentration of a solution.

11. A buffer helps to minimize changes in pH.

Sections 2.5, 2.6, 2.7, 2.8, 2.9, 2.10

12. Organic molecules are molecules that contain carbon covalently bonded to other elements.

13. Dehydration synthesis creates macromolecules by removing water. Hydrolysis breaks down macromolecules by adding water.

14. The four classes of organic molecules synthesized by living organisms are carbohydrates, lipids, proteins, and nucleic acids.

15. Carbohydrates in the form of simple sugars provide immediate energy to the cells, while polysaccharides provide stored energy.

16. Lipids include triglycerides that store energy, phospholipids that function in the cell membrane, and steroids that are classified as lipids because they are not soluble in water.

17. Proteins are composed of amino acids and are described at four structural levels.

18. Enzymes are proteins that function as catalysts to speed up chemical reactions.

19. Nucleic acids are composed of nucleotides.

20. DNA and RNA are nucleic acids that function in storing and expressing genetic information. ATP is a nucleic acid that functions as an energy molecule in the cell.

Exercises

Complete the exercises for each section after you have read and studied the section. If you cannot answer some questions, or answer them incorrectly, return to the chapter and review this information. You may find it helpful to work on only one section at a time. When you have completed all sections, take the Chapter Test as an indicator of your mastery of this topic.

2.1 All matter consists of elements

2.2 Atoms combine to form molecules

Matching

_____ 1. **matter** a. atoms of an element with the same atomic number but a different atomic mass

_____ 2. **element** b. stored energy that is not currently doing work

_____ 3. **atom** c. a weak attractive force between oppositely charged regions of polar molecules containing hydrogen

_____ 4. **protons** d. anything that has mass and occupies space

_____ 5. **neutrons** e. enzymes and nutrients that protect the body from the action of free radicals

continued

_____ 6. **electron**   f. a pure form of matter that cannot be broken down to a simpler form

_____ 7. **isotope**   g. energy that is doing work, the energy of motion

_____ 8. **molecule**   h. a bond formed by the sharing of electrons

_____ 9. **energy**   i. the smallest unit of any element that still retains the physical and chemical properties of that element

_____ 10. **potential energy**   j. attractive forces that bind atoms together

_____ 11. **kinetic energy**   k. neutral particles found in the nucleus of an atom

_____ 12. **chemical bonds**   l. the area within a shell where an electron is likely to be found

_____ 13. **covalent bond**   m. positively charged particles found in the nucleus of an atom

_____ 14. **ion**   n. the capacity to do work

_____ 15. **ionic bond**   o. negatively charged particles found orbiting around the nucleus of an atom

_____ 16. **hydrogen bond**   p. a stable association between two or more atoms

_____ 17. **orbital**   q. a bond formed by the attraction of two oppositely charged ions

_____ 18. **antioxidant**   r. an electrically charged atom

Fill-in-the-Blank

Referenced sections are indicated in parentheses.

19. The central core of an atom is called the _____. (2.1)

20. The second shell of an atom can hold up to _____ electrons. (2.1)

21. The _____ _____ represents the number of protons in an atom's nucleus. (2.1)

22. The _____ _____ represents the total number of neutrons and protons in the nucleus. (2.1)

23. All atoms of the same element have the same number of _____. (2.1)

24. _____ are unstable isotopes that give off energy in the form of radiation. (2.1)

25. Chemical bonds store _____ energy. (2.2)

26. Electrons occupy an area within a shell called a(n) _____. (2.2)

27. Ions in solution may be called _____. (2.2)

28. _____ molecules are electrically neutral overall, but they have partially charged ends. (2.2)

29. The six most common elements in the body are _____, _____, _____, _____, _____, and _____. (2.2)

30. _____ bonds are formed when atoms share two pairs of electrons. (2.2)

31. The attraction of two oppositely charged ions results in a(n) _____ bond. (2.2)

32. A weak attraction between oppositely charged regions of polar molecules containing H is a _____ bond. (2.2)

Labeling

Use the terms listed to label Figure 2.1.

a. nucleus
b. proton
c. electron
d. neutron
e. first shell
f. second shell

Figure 2.1

2.3 Life depends on water

2.4 The importance of hydrogen ions

Matching

___ 1. **solvent** a. a measure of H⁺ concentration in a solution

___ 2. **acid** b. a liquid in which other substances dissolve

___ 3. **base** c. any substance that minimizes changes in the pH of a solution when an acid or a base is added

___ 4. **pH scale** d. a molecule that can accept a hydrogen ion

___ 5. **buffer** e. a molecule that can give up a hydrogen ion

Completion

Indicate which words or phrases below correspond to Figure 2.2.

a. acidic solution

b. alkaline solution

c. neutral solution

d. higher H⁺ concentration than pure water

e. same H⁺ concentration as pure water

f. lower H⁺ concentration than pure water

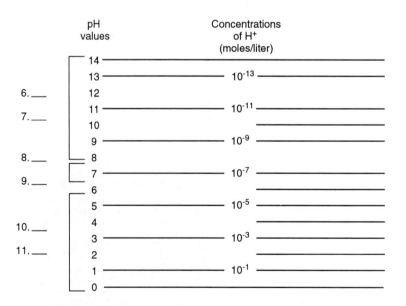

Figure 2.2

2.5 The organic molecules of living organisms
2.6 Carbohydrates: Used for energy and structural support
2.7 Lipids: Insoluble in water

Matching

____ 1. **organic molecule**   a. a long straight or branched chain containing thousands of monosaccharides

____ 2. **dehydration synthesis**   b. a polysaccharide used by plants for structural support

____ 3. **hydrolysis**   c. the class of biochemical compounds that contains triglycerides, phospholipids, and steroids

____ 4. **monosaccharide**   d. a group of molecules classified as lipids because they are insoluble in water

____ 5. **oligosaccharide**   e. a molecule that contains carbon covalently bonded to other elements

____ 6. **polysaccharide**   f. a steroid that forms part of the cell membrane

____ 7. **glycogen**   g. a reaction in which small molecules are covalently bonded by the removal of a water molecule

____ 8. **starch**   h. a lipid composed of one molecule of glycerol and three fatty acids

____ 9. **cellulose**   i. a reaction in which macromolecules are broken apart by the addition of a water molecule

____ 10. **lipid**   j. a lipid that forms an important part of the cell membrane

____ 11. **triglyceride**   k. a short string of monosaccharides linked together

____ 12. **fatty acid**   l. the storage polysaccharide found in plants

____ 13. **phospholipid**   m. the simplest carbohydrate, consisting of a single sugar unit

____ 14. **steroid**   n. the storage polysaccharide found in animals

____ 15. **cholesterol**   o. a chain of hydrocarbons with a terminal carboxyl group

Fill-in-the-Blank

Referenced sections are indicated in parentheses.

16. _____ are large organic molecules consisting of thousands of smaller molecules. (2.5)

17. The synthesis of macromolecules from smaller molecules _____ energy. (2.5)

18. The breakdown of macromolecules _____ energy. (2.5)

continued

19. The four classes of organic molecules made by living organisms are _____, _____, _____, and _____. (2.5)

20. _____ is a six-carbon monosaccharide that serves as an important source of energy for cells. (2.6)

Labeling

Use the words below to identify each molecule or molecular region indicated in Figure 2.3.

a. glycerol
b. steroid
c. monosaccharide
d. fatty acids
e. unsaturated fatty acid
f. triglyceride
g. disaccharide
h. polar head
i. saturated fatty acid
j. phospholipid
k. fatty acid tails

Figure 2.3

2.8 Proteins: Complex structures constructed of amino acids

2.9 Nucleic acids store genetic information

2.10 ATP carries energy

Matching

____ 1. **amino acids**   a. single units joined by covalent bonds in DNA and/or RNA

____ 2. **polypeptide**   b. a protein that speeds up a chemical reaction

____ 3. **protein**   c. a nucleotide that functions as the universal energy source for cells

____ 4. **denaturation**   d. single units joined by covalent bonds to form a protein

____ 5. **enzyme**   e. a chain of more than 100 amino acids

____ 6. **DNA, RNA**   f. permanent damage to protein structure resulting in loss of function

____ 7. **nucleotides**   g. a string of 3–100 amino acids

____ 8. **ATP**   h. nucleic acids

Fill-in-the-Blank

Referenced sections are indicated in parentheses.

9. An enzyme can facilitate a reaction in which two reactants are joined to form a _____. (2.8)

10. A nucleotide consists of a _____, a _____, and a _____. (2.9)

11. DNA contains the instructions for producing _____. (2.9)

12. RNA contains the instructions for producing _____. (2.9)

13. The four bases found in DNA nucleotides are _____, _____, _____, and _____. (2.9)

14. RNA is a _____-stranded molecule containing a _____ sugar. (2.9)

15. The breakdown of ATP produces _____, _____, and _____. (2.10)

Paragraph Completion

Use the following terms to complete the paragraph, and then repeat this exercise with the terms covered.

beta-pleated sheet	denaturation	covalent	tertiary
amino	primary	temperature	R
carboxylic acid	secondary	amino acids	disulfide
structure	quaternary	pH	polar

Proteins are macromolecules consisting of long chains of (16) _____ joined by (17) _____ bonds. Each amino acid has a central C-H group, bonded to an (18) _____ group, a (19) _____ group, and an additional (20) _____ group representing one of 20 different chemical possibilities. Protein function depends on the protein's (21) _____. The (22) _____ structure of a protein is the sequence of amino acids. Hydrogen bonding between amino acids at regular intervals produces the (23) _____ structure. Examples of this structure are the alpha-helix and the (24) _____. Twisting and folding of a protein into a three-dimensional shape produces the (25) _____ structure. This is the result of hydrogen bonds and (26) _____ bonds. Some proteins also have a (27) _____ structure, produced when two or more polypeptides are associated with each other. Because protein structure is maintained by weak hydrogen bonds, proteins can temporarily change their shape in response to (28) _____ molecules. A change in protein shape that leads to loss of function is called (29) _____ and may be caused by high (30) _____ or changes in (31) _____.

Labeling

Use the terms below to identify each molecule or molecular region indicated in Figure 2.4.

a. triphosphate	c. R group	e. ribose	g. ATP
b. amino group	d. adenine	f. carboxylic acid group	h. amino acid

Figure 2.4

Chapter Test

Multiple Choice

1. Which of the following statements about matter is false?
   a. Matter is anything that has mass and occupies space.
   b. Matter is composed of elements.
   c. An element is a pure form of matter that cannot be broken down to a simpler form.
   d. The periodic table of elements arranges all known elements according to their physical chemical properties.

2. Protons carry _____ charge and are located _____.
   a. a negative, outside the nucleus
   b. no, in the nucleus
   c. a positive, outside the nucleus
   d. a positive, in the nucleus

3. The first electron shell can hold _____ electrons, and the second shell can hold _____ electrons.
   a. 2, 2
   b. 2, 8
   c. 8, 8
   d. 2, 10

4. Sodium has an atomic number of 11. This indicates that sodium has:
   a. 11 neutrons.
   b. a total of 11 particles in the nucleus.
   c. 11 protons.
   d. a total of 11 particles in the entire atom.

5. Most atoms of hydrogen have no neutrons. An atom of hydrogen containing one neutron would:
   a. be an ion.
   b. be an isotope.
   c. be a molecule.
   d. also have an additional proton.

6. Kinetic energy is:
   a. potential energy.
   b. unusable energy.
   c. inaccessible energy.
   d. the energy of motion.

7. Electrons with the highest potential energy:
   a. are found closest to the nucleus.
   b. are found farthest from the nucleus.
   c. are found in the nucleus.
   d. are bound to protons.

8. Atoms are most stable when:
   a. they have the same number of protons as neutrons.
   b. they have the same number of protons as electrons.
   c. their outermost occupied electron shell is partially filled.
   d. their outermost occupied electron shell is full.

9. Atoms that are not stable tend to:
   a. form chemical bonds with other atoms.
   b. repel other atoms.
   c. be isolated from other atoms.
   d. have higher kinetic energy.

10. Covalent bonds involve _____ electrons, and ionic bonds involve _____ electrons.
    a. transferring, sharing
    b. transferring, transferring
    c. sharing, transferring
    d. sharing, sharing

11. An electrically charged atom or molecule is called a(n):
    a. ion.
    b. isotope.
    c. compound.
    d. organic.

12. When an atom loses an electron, it becomes:
    a. positively charged.
    b. negatively charged.
    c. neutral.
    d. an isotope.

13. Water molecules are bonded to other water molecules by:
    a. covalent bonds.
    b. ionic bonds.
    c. hydrogen bonds.
    d. water bonds.

14. Water is a good biological solvent because:
    a. it is a polar molecule.
    b. it is a liquid at body temperature.
    c. it forms covalent bonds with dissolved substances.
    d. a and b.

15. An acid is a molecule that _____ H^+, while a base is a molecule that _____ H^+.
    a. absorbs, releases
    b. releases, absorbs
    c. reacts with, does not react with
    d. decreases, increases

16. Blood pH is maintained by the carbonic acid/bicarbonate buffer system: $HCO_3^- + H^+ \leftrightarrow H_2CO_3$. If the blood becomes too acidic, which of the following will occur?
    a. Carbonic acid will release H^+.
    b. Carbonic acid will absorb H^+.
    c. Bicarbonate will release H^+.
    d. Bicarbonate will absorb H^+.

17. The synthesis of macromolecules from smaller molecules:
    a. releases energy.
    b. requires energy.
    c. is accomplished by hydrolysis.
    d. requires the addition of water.

18. Monosaccharides:
    a. contain C, H, and O in a 1–2–1 ratio.
    b. contain C and H in a 1–1 ratio.
    c. contain equal amounts of C and H, with a small amount of O.
    d. contain equal amounts of C and O, with a small amount of H.

19. Which of the following is a carbohydrate with structural importance in plants?
    a. glycogen
    b. glucose
    c. starch
    d. cellulose

20. Triglycerides with a full complement of two H atoms for each C atom in their fatty-acid tails are:
    a. phospholipids.
    b. steroids.
    c. saturated.
    d. unsaturated.

21. Proteins function as catalysts when they:
    a. act as enzymes.
    b. speed up the rate of a chemical reaction.
    c. store genetic information.
    d. both a and b

22. The three-dimensional protein shape maintained by hydrogen and disulfide bonds is the:
    a. primary structure.
    b. secondary structure.
    c. tertiary structure.
    d. quaternary structure.

23. What molecule is composed of a sugar, a phosphate, and a base?
    a. an amino acid
    b. a nucleotide
    c. a quaternary protein
    d. a saturated fat

24. Which of the following is *not* different in RNA and DNA?
    a. the phosphate group
    b. the sugar
    c. the presence of uracil
    d. the number of strands

25. The breakdown of ATP produces:
    a. ADP.
    b. free phosphate.
    c. energy.
    d. all of the above.

Key Concept Review Questions

Each of the Key Concepts listed at the beginning of this chapter has been rewritten as a question below. After successfully completing the study guide exercises and the Chapter Test, you should be able to answer each of these questions. Refer to the Key Concepts list at the beginning of this chapter to check your answers.

1. What is matter composed of? What are elements composed of? What are atoms composed of?
2. What is indicated by the atomic number? What is indicated by the mass number?
3. What is an isotope?
4. What is energy? What type of energy is possessed by electrons based on their position?
5. What atoms tend to form chemical bonds?
6. Name three main types of bonds and describe how each is formed.
7. What characteristics of water make it a good biological solvent?
8. How does water help to regulate body temperature?
9. What is an acid? What is a base?
10. What is the pH scale?
11. What is the function of a buffer?
12. How can organic molecules be described?
13. How does dehydration synthesis differ from hydrolysis?
14. What are the four classes of organic molecules synthesized by living organisms?
15. Simple sugars and polysaccharides are both involved in energy; how are their roles different?
16. What are the three types of lipids and their functions?
17. What are proteins composed of? How are proteins characterized?
18. What is an enzyme?
19. What are nucleic acids composed of?
20. What are the functions of DNA, RNA, and ATP?

3

Structure and Function of Cells

Chapter Summary and Key Concepts

After reading and studying this chapter you should know the following:

Sections 3.1, 3.2, 3.3, 3.4

1. The cell doctrine states that all living things are composed of cells, a cell is the smallest unit of life, and all cells come from preexisting cells.

2. Eukaryotic cells have a membrane-bound nucleus; prokaryotic cells do not.

3. Eukaryotic cells have a plasma membrane, a nucleus, and cytoplasm.

4. Cells remain small to stay efficient.

5. The plasma membrane is selectively permeable and consists of a lipid bilayer, proteins, cholesterol, and carbohydrates that extend from the outer surface.

6. Passive transport moves molecules with the gradient and requires no energy; active transport moves molecules against the gradient and requires energy.

7. Membrane proteins are required for some processes of diffusion, facilitated diffusion, and active transport.

8. Osmosis is the net diffusion of water across a selectively permeable membrane.

9. Endocytosis and exocytosis move materials in bulk. Endocytosis moves material into the cell, and exocytosis moves material out of the cell.

10. Receptor proteins receive and transmit information across the plasma membrane.

11. The sodium-potassium pump is an important active transport system that helps to maintain ion gradients and cell volume.

Sections 3.5, 3.6

12. The nucleus is surrounded by a nuclear membrane consisting of two lipid bilayers studded with pores.

13. Ribosomes synthesize proteins and may be free in the cytoplasm or bound to the rough endoplasmic reticulum.

14. The smooth endoplasmic reticulum synthesizes nonprotein molecules and packages proteins in vesicles for movement to the Golgi apparatus.

15. The Golgi apparatus refines proteins and packages them to be shipped.

16. Vesicles are membrane-bound sacs that enclose substances within the cell. They may be secretory, endocytotic, for shipping and storage, or lysosomal.

17. Mitochondria harvest the energy from ingested nutrients and use it to synthesize ATP.

18. Cell structures that contain microtubules include the cytoskeleton which provides structural support, cilia and flagella that aid movement, and centrioles that participate in cell division.

Section 3.7

19. Cells use and transform matter and energy through chemical reactions. Metabolism is the sum of all chemical reactions in an organism.

20. Cellular respiration breaks down glucose in the presence of oxygen to yield 36 molecules of ATP. In the absence of oxygen, a cell can produce two molecules of ATP using anaerobic pathways.

Exercises

Complete the exercises for each section after you have read and studied the section. If you cannot answer some questions, or answer them incorrectly, return to the chapter and review this information. You may find it helpful to work on only one section at a time. When you have completed all sections, take the Chapter Test as an indicator of your mastery of this topic.

3.1 Cells are classified according to their internal organization

3.2 Cell structure reflects cell function

3.3 A plasma membrane surrounds the cell

3.4 Molecules cross the plasma membrane in several ways

Matching

___ 1. **cell doctrine**
___ 2. **plasma membrane**
___ 3. **eukaryotes**
___ 4. **cytoplasm**
___ 5. **organelle**
___ 6. **prokaryotes**
___ 7. **passive transport**
___ 8. **diffusion**
___ 9. **selective permeability**
___ 10. **osmosis**
___ 11. **facilitated transport**
___ 12. **active transport**
___ 13. **sodium-potassium pump**
___ 14. **exocytosis**
___ 15. **endocytosis**
___ 16. **receptor protein**
___ 17. **isotonic solution**
___ 18. **hypertonic solution**
___ 19. **hypotonic solution**

a. the movement of molecules from one area to another as a result of random motion
b. material within a cell, composed of fluid and organelles
c. a solution with a concentration of solutes higher than the intracellular fluid
d. diffusion of substances through the plasma membrane with the help of transport proteins
e. three general principles that describe cells and their relationship to living things
f. movement of a substance through the plasma membrane against the concentration gradient
g. movement of material out of the cell by fusion of a vesicle with the plasma membrane
h. the outer membrane that surrounds all cells
i. the nature of the plasma membrane to allow some substances to cross by diffusion, but not others
j. an active transport pump important in maintaining cell volume
k. movement of materials into the cell by enclosure in a vesicle formed by the plasma membrane
l. membrane proteins with specific receptor sites that can trigger biochemical events within the cell
m. organisms composed of cells that have a membrane-bound nucleus
n. same solute concentration in the extracellular fluid and the intracellular fluid
o. the net diffusion of water across a selectively permeable membrane
p. transport of a molecule through the plasma membrane without the use of energy
q. organisms whose cells do not have a membrane-bound nucleus
r. a solution with a concentration of solutes lower than the intracellular fluid
s. structure within a cell that carries out specialized functions

Fill-in-the-Blank

Referenced sections are indicated in parentheses.

20. Eukaryotic cells contain three basic structural components, the _____, _____, and _____ _____. (3.1)

21. Some cells increase their surface area without significantly increasing their volume with a projection of the cell membrane called _____. (3.2)

22. The _____ microscope bombards a sample with a beam of electrons to produce a two-dimensional image. (3.2)

23. The plasma membrane is composed of two layers of phospholipids called the _____ _____. (3.3)

24. The changing structure and protein pattern of the plasma membrane is referred to by the term _____ _____. (3.3)

25. A _____ exists when there is a difference in the concentration of a solution in different areas. (3.4)

26. _____ exists when molecules are diffusing randomly, but equally, in all directions. (3.4)

27. The fluid pressure required to oppose osmosis is called _____ _____. (3.4)

28. When ATP is used to power active transport it will break down to produce _____, _____, and _____. (3.4)

29. The area outside a cell is referred to as the _____ environment. (3.4)

30. When the sodium-potassium pump is being used to reduce cell volume, the _____ of the pump will increase. (3.4)

Labeling

Use the terms below to label Figure 3.1.

a. phospholipid
b. receptor protein
c. lipid bilayer
d. gated protein
e. cytoskeleton filaments
f. transport protein
g. open protein channel
h. glycoprotein
i. cholesterol
j. carbohydrate groups

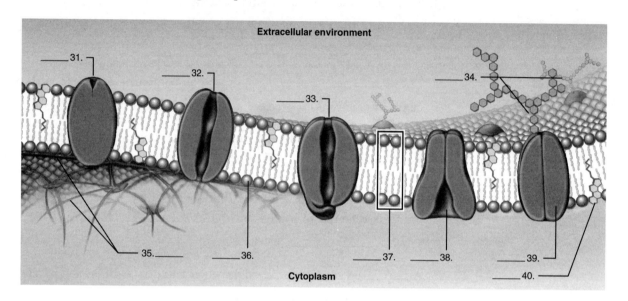

Figure 3.1

Completion

*For each of the membrane transport processes listed below, indicate **all** of the following characteristics that apply in the table on page 33 (characteristics may be used more than once).*

a. requires energy
b. does not require energy
c. moves a substance through the lipid bilayer of the plasma membrane
d. moves a substance through a protein channel
e. moves a substance with the help of a transport protein
f. moves a substance with the gradient
g. moves a substance against the gradient
h. moves a substance within a vesicle
i. moves water only

Transport Process	Characteristics
41. Diffusion through the lipid bilayer	
42. Diffusion through protein channels	
43. Facilitated transport	
44. Osmosis	
45. Active transport	
46. Endocytosis	
47. Exocytosis	

Word Choice

Circle the accurate term or phrase to complete each sentence.

48. Cells must maintain a (small/large) size to be efficient. (3.2)

49. The concentration of water in a solution is (the same as/opposite to) the concentration of solutes in the solution. (3.4)

50. Polar molecules (can/cannot) cross the lipid bilayer portion of the cell membrane. (3.4)

51. When a substance moves against the gradient, it moves from an area of (low/high) concentration to an area of (low/high) concentration. (3.4)

52. Gated protein channels are normally (open/closed) and will (open/close) under certain conditions. (3.4)

53. Moving a substance with the gradient (does/does not) require energy. (3.4)

54. The sodium-potassium pump moves three sodium ions (into/out of) the cell and two potassium ions (into/out of) the cell. (3.4)

55. In a hypertonic environment, water will move (into/out of) the cell. (3.4)

3.5 Internal structures carry out specific functions

3.6 Cells have structures for support and movement

Crossword Puzzle

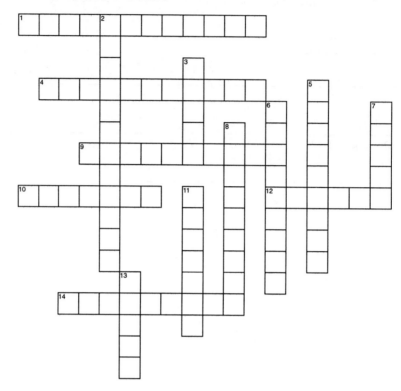

Across

1. Synthesizes ATP
4. Vesicles containing enzymes that detoxify wastes
9. Rodlike structures composed of microtubules that aid the process of cell division
10. An organelle that contains the cell's genetic material
12. The _____ endoplasmic reticulum synthesizes lipids
14. The endoplasmic _____ is a membranous tunnel-like system running through the cytoplasm

Down

2. A network of microtubules and microfilaments that provide structural support to the cell
3. The _____ apparatus refines proteins and packages them to be shipped to their final destination
5. Responsible for protein synthesis in the cytoplasm
6. Vesicles containing powerful digestive enzymes
7. The _____ endoplasmic reticulum has attached ribosomes
8. Long whiplike projection, found in humans only on sperm cells
11. A membrane-bound sphere that encloses something within the cell
13. Short hairlike projections found covering a cell surface

Labeling

Label the indicated structures in Figure 3.2.

Figure 3.2

3.7 Cells use and transform matter and energy

Matching

___	1. **metabolism**	a.	electron transport molecules
___	2. **metabolic pathway**	b.	the starting material in a chemical reaction
___	3. **substrate**	c.	a metabolic pathway in which large molecules are broken down into smaller molecules
___	4. **product**	d.	production of ATP in the absence of oxygen
___	5. **anabolism**	e.	the sum of all the chemical reactions in an organism
___	6. **catabolism**	f.	metabolic processes within a cell that utilize oxygen and produce carbon dioxide in the process of making ATP
___	7. **cellular respiration**	g.	the ending material in a chemical reaction
___	8. **NAD$^+$, FAD**	h.	a metabolic pathway in which small molecules are assembled into larger molecules
___	9. **anaerobic pathway**	i.	a series of reactions following one another in an orderly and predictable manner.

Paragraph Completion

Use the following terms to complete the paragraphs, and then repeat this exercise with the terms covered. Terms used more than once are listed multiple times.

hydrogen	acetyl groups	$FADH_2$	cytoplasm
ATP synthase	carbohydrates	pyruvate	lipids
mitochondria	acetyl groups	pyruvate	proteins
acetyl CoA	two	coenzyme A	NADH
cyclic	inner	hydrogen	two
active transport	gradient	energy	inner compartment

The production of ATP occurs primarily in the (10) _____ and involves the catabolism of (11) _____, (12) _____, and (13) _____. Cellular respiration often begins with a molecule of glucose and proceeds through the following stages:

Glycolysis: Glycolysis occurs in the (14) _____. During glycolysis, a 6-C glucose molecule is broken down into two 3-C molecules called (15) _____. A net yield of (16) _____ ATP occurs.

Preparatory Step: In this preparatory step for the citric acid cycle, the two (17) _____ molecules produced in glycolysis enter the mitochondria and are converted into two molecules of (18) _____. Only the (19) _____ actually enter the citric acid cycle; (20) _____ acts as a shuttle.

Citric Acid Cycle: The citric acid cycle is an example of a (21) _____ pathway. During these chemical reactions, the (22) _____ are broken down and (23) _____ ions and electrons are transferred to NAD^+ and FAD, forming (24) _____ and (25) _____. (26) _____ molecules of ATP are also generated by the citric acid cycle; however, most of the energy of glucose is still trapped in the (27) _____ being carried by NADH and $FADH_2$.

Electron Transport System: The electron transport system is located on the (28) _____ membrane of the mitochondria. Hydrogen electrons delivered by NADH and $FADH_2$ are accepted by the proteins of the electron transport system, and release (29) _____ as they move from one transport molecule to the next. This energy is used for the (30) _____ of H⁺ into the outer compartment, creating a H⁺ (31) _____. H⁺ are allowed to flow back into the (32) _____ only through a special protein channel that functions as an enzyme called (33) _____. This enzyme uses the energy released by the flowing H⁺ to synthesize ATP.

Short Answer

34. What happens to pyruvate when oxygen is not available to the cell?

35. How does substrate level phosphorylation differ from oxidative phosphorylation?

36. What is the final fate of low energy H⁺ and electrons?

37. Why do fats store more than twice the energy of carbohydrates?

Chapter Test

Multiple Choice

1. Which of the following is not part of the cell doctrine?
    a. All living things are composed of cells.
    b. Cells must remain small to function effectively.
    c. Cells are the smallest units of life.
    d. All cells come from preexisting cells.

2. You suspect a cell to be prokaryotic. Which of the following would confirm this?
    a. The cell lacks a plasma membrane.
    b. The cell lacks a nuclear membrane.
    c. The cell lacks genetic material.
    d. The cell lacks cytoplasm.

3. The plasma membrane is composed of lipids arranged in a:
    a. single layer with the hydrophilic heads near the extracellular fluid.
    b. single layer with the hydrophobic tails near the extracellular fluid.
    c. bilayer with the tails facing the intracellular and extracellular fluid.
    d. bilayer with the heads facing the intracellular and extracellular fluid.

4. Cholesterol in the plasma membrane functions:
   a. as a receptor molecule.
   b. as a transport molecule.
   c. to make the membrane more rigid.
   d. to make the membrane more permeable.

5. The maintenance of cell shape is least affected by:
   a. the stiffness of the plasma membrane.
   b. the network of cytoskeleton protein fibers.
   c. the intracellular fluid.
   d. limitations imposed by surrounding cells.

6. The "fluid mosaic" model of the plasma membrane refers to:
   a. the ability of proteins to drift about in the membrane.
   b. the fluidity of the lipid bilayer.
   c. the presence of organelles in the cytoplasm.
   d. a and b.
   e. all of the above.

7. Passive transport:
   a. requires energy.
   b. relies on the process of diffusion.
   c. is dependent on active transport.
   d. moves a substance against the gradient.

8. Which of the following move bulk substances into the cell by forming a membrane-bound vesicle?
   a. passive transport
   b. osmosis
   c. endocytosis
   d. exocytosis

9. Oxygen and carbon dioxide move into and out of the cell by:
   a. diffusion through the lipid bilayer.
   b. facilitated diffusion.
   c. active transport.
   d. osmosis.

10. Substance X is more concentrated in the intracellular fluid than the extracellular fluid. Substance X will move into the cell by:
    a. diffusion.
    b. facilitated diffusion.
    c. exocytosis.
    d. active transport.

11. How does facilitated diffusion differ from active transport?
    a. Facilitated diffusion requires transport proteins.
    b. Facilitated diffusion does not require energy.
    c. Facilitated diffusion moves a substance down the concentration gradient.
    d. Both b and c.
    e. All of the above.

12. The sodium-potassium pump is important in maintaining cell volume because:
    a. it transports water into and out of the cell.
    b. it blocks the movement of water with membrane proteins.
    c. it transports ions, which results in the movement of water by osmosis.
    d. it counteracts the effect of osmotic pressure.

13. Which of the following events may contribute to a transfer of information across the plasma membrane?
    a. receptor proteins receive and transmit information
    b. a molecule binds a receptor site on a receptor protein
    c. endocytosis brings receptor proteins into the cell
    d. both a and b
    e. all of the above

14. A cell placed in an isotonic environment will:
    a. swell.
    b. shrink.
    c. stay the same size.
    d. open protein channels.

15. The nucleolus is:
    a. an area of concentrated DNA.
    b. an area of RNA synthesis.
    c. an area outside the nucleus where ribosomes are assembled.
    d. a small additional nucleus found in large cells.

16. A cell that lacked rough endoplasmic reticulum (ER) would be unable to:
    a. synthesize lipids.
    b. detoxify wastes.
    c. manufacture ATP.
    d. synthesize proteins for use outside the cell.

17. Vesicles fusing with the Golgi apparatus would have arisen from the:
    a. plasma membrane.
    b. smooth ER.
    c. mitochondria.
    d. nucleus.

18. Vesicles fusing with the plasma membrane in the process of _____ would have arisen from the _____.
    a. exocytosis, Golgi apparatus
    b. endocytosis, Golgi apparatus
    c. exocytosis, rough ER
    d. endocytosis, nucleus

19. Cellular debris will be removed by:
    a. peroxisomes.
    b. endocytotic vesicles.
    c. secretory vesicles.
    d. lysosomes.

20. Which of the following is composed of a circle of nine microtubule pairs surrounding a single pair in the center?
    a. centrioles
    b. cilia
    c. mitochondria
    d. ribosomes

21. Glycolysis is all of the following *except:*
    a. aerobic.
    b. a linear pathway.
    c. a process that occurs in the cytoplasm.
    d. a process that produces ATP.

22. The breakdown of macromolecules is a(n) _____ process and _____ energy.
    a. anabolic, requires
    b. anabolic, releases
    c. catabolic, requires
    d. catabolic, releases

23. Which of the following stages of cellular respiration does not contain any carbon atoms from the original glucose molecule?
    a. citric acid cycle
    b. electron transport system
    c. glycolysis
    d. formation of acetyl CoA

24. How much ATP could a cell produce if no molecules of NAD^+ and FAD were available?
    a. 0
    b. 4
    c. 12
    d. 32

25. The complete catabolism of glucose in the presence of oxygen produces _____ molecules of ATP, while anaerobic pathways produce _____ molecules of ATP.
    a. 32, 4
    b. 32, 2
    c. 36, 4
    d. 36, 2

Key Concept Review Questions

Each of the Key Concepts listed at the beginning of this chapter has been rewritten as a question below. After successively completing the study guide exercises and the Chapter Test, you should be able to answer each of these questions. Refer to the Key Concepts list at the beginning of this chapter to check your answers.

1. What three principles are stated by the cell doctrine?
2. How are eukaryotic cells different from prokaryotic cells?
3. What three basic components are found in eukaryotic cells?
4. Why are all cells small?
5. How permeable is the plasma membrane? What components are found as part of the plasma membrane?
6. How does passive transport differ from active transport?
7. What transport processes require membrane proteins?
8. What is osmosis?
9. What occurs in endocytosis and exocytosis?
10. What is the role of receptor proteins?
11. Why is the sodium-potassium pump important to the cell?
12. How would you describe the structure of the nuclear membrane?
13. What is the function of ribosomes? Where are ribosomes located?
14. What is the function of the smooth endoplasmic reticulum?
15. What is the function of the Golgi apparatus?
16. What are vesicles? List four different types of vesicles.
17. What is the function of mitochondria?
18. List four cell structures that contain microtubules, and describe the function of each.
19. How do cells use and transform energy? What is metabolism?
20. What molecules are involved in the process of cellular respiration? How much ATP is produced? How is this process different in the absence of oxygen?

4

From Cells To Organ Systems

Chapter Summary and Key Concepts

After reading and studying this chapter you should know the following:

Sections 4.1, 4.2, 4.3

1. Multicellular organisms consist of many cells, allowing them to achieve greater size and greater ability to maintain a desired environment.

2. Tissues are groups of cells that perform common functions. The four major tissue types of the body are epithelial, connective, muscle, and nervous tissue.

3. Epithelial tissue covers or lines body surfaces or cavities, and is classified by the number of cell layers and cell shape.

4. Glands are a type of epithelial tissue specialized to secrete a product.

5. A basement membrane anchors epithelial tissue to the underlying connective tissue. Junctions anchor epithelial cells to other cells.

6. Connective tissue supports soft organs and connects parts of the body.

7. Connective tissue consists of a nonliving *matrix* of background substance and fibers, and cells responsible for secreting the matrix.

8. Connective tissue can be classified as fibrous or specialized.

9. Fibrous connective tissue includes the following types: loose, dense, elastic, and reticular.

10. Specialized connective tissue includes the following types: cartilage, bone, blood, and adipose tissue.

11. Each type of connective tissue will vary in the type of matrix, cells, and fibers it contains. The structure of the tissue allows it to perform its function.

Sections 4.4, 4.5

12. Muscle tissue consists of cells specialized to produce movement. Muscle tissue types include skeletal, smooth, and cardiac.

13. Skeletal muscle cells are voluntary, multinucleated, and attached to bones.

14. Smooth muscle cells are involuntary, have a single nucleus, and are found surrounding hollow organs and tubes.

15. Cardiac muscle cells are involuntary, have a single nucleus, and are found in the heart.

16. Nervous tissue consists of cells specialized to generate and transmit electrical impulses used for communication in the body. Nervous tissue cells include neurons responsible for communication, and glial cells that support the neurons.

Sections 4.6, 4.7, 4.8

17. Organ systems are composed of groups of organs that perform a broad function. The human body has 11 organ systems.

18. Tissue membranes line body cavities. The four major types of tissue membranes are serous, mucous, synovial, and cutaneous.

19. The integumentary system consists of the epidermis and the dermis, and it functions in protection from dehydration, injury, and invasion; regulation of body temperature; synthesis; and sensation.

20. Multicellular organisms maintain homeostasis using negative feedback control systems. These systems include a controlled variable, a sensor, a control center, and an effector.

Exercises

Complete the exercises for each section after you have read and studied the section. If you cannot answer some questions, or answer them incorrectly, return to the chapter and review this information. You may find it helpful to work on only one section at a time. When you have completed all sections, take the Chapter Test as an indicator of your mastery of this topic.

4.1 Tissues are groups of cells with a common function

4.2 Epithelial tissues cover body cavities and surfaces

4.3 Connective tissue supports and connects body parts

Matching

____ 1. **tissue**
____ 2. **epithelial tissue**
____ 3. **glands**
____ 4. **basement membrane**
____ 5. **connective tissue**
____ 6. **collagen fibers**
____ 7. **elastic fibers**
____ 8. **reticular fibers**
____ 9. **fibroblasts**
____ 10. **cartilage**
____ 11. **bone**
____ 12. **blood**
____ 13. **adipose tissue**
____ 14. **adipocytes**
____ 15. **tight junction**
____ 16. **adhesion junction**
____ 17. **gap junction**

a. a connective tissue composed primarily of collagen fibers, with chondroblasts in lacunae
b. fat cells
c. connective tissue protein fibers that allow a tissue to stretch
d. a connective tissue specialized for fat storage
e. prevents the passage of materials between two cells
f. a group of specialized cells that perform a common function
g. a connective tissue consisting of cells suspended in a fluid called plasma
h. sheets of cells that line or cover body surfaces or cavities
i. connective tissue cells that secrete protein fibers
j. anchors two cells together while allowing flexibility
k. a tissue that supports and connects parts of the body
l. epithelial tissue specialized to secrete a product
m. connective tissue fibers that support or provide an internal framework for a body organ
n. connecting channels between two cells that permit the passage of ions
o. connective tissue protein fibers that confer strength to a tissue
p. a supporting layer beneath epithelial tissue that anchors it to connective tissue
q. a hard connective tissue containing calcium and phosphate

Labeling

Label each epithelial tissue in Figure 4.1 with the (a) tissue type and (b) function.

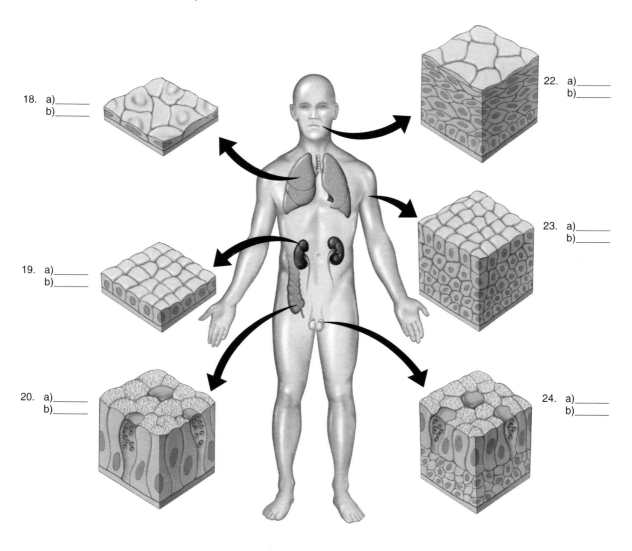

18. a)_____
    b)_____

19. a)_____
    b)_____

20. a)_____
    b)_____

22. a)_____
    b)_____

23. a)_____
    b)_____

24. a)_____
    b)_____

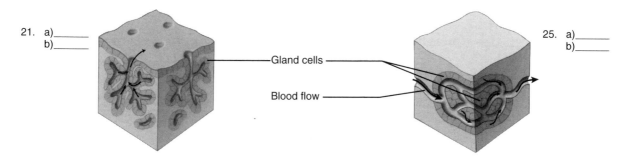

21. a)_____
    b)_____

25. a)_____
    b)_____

Gland cells
Blood flow

Figure 4.1

Paragraph Completion

Use the following terms to complete the paragraphs, then repeat this exercise with the terms covered. Terms used more than once are listed multiple times.

collagen	reticular	fibers	chondroblasts
connective	matrix	special	reticular
blood	gel-like	elastic	adipocytes
collagen	dense	loose	collagen
fat	elastic	solid	fibroblasts

(26) _____ tissue supports and connects various parts of the body. Most connective tissues consist of a few cells scattered in a nonliving extracellular material called the (27) _____. The matrix usually consists of protein (28) _____ embedded in a material that is liquid, gel-like, or solid. Connective tissues are broadly categorized as fibrous or (29) _____.

Fibrous connective tissue has fibers and cells embedded in a (30) _____ ground substance. The three types of fibers are (31) _____ fibers, which add strength; (32) _____ fibers, which help a tissue to stretch without breaking; and (33) _____ fibers, which often form an internal framework for an organ. The most important cell type present is a (34) _____, which secrete the fibers of the matrix. Fibrous connective tissues are further classified according to the type and arrangement of fibers. The most common type is (35) _____ connective tissue which is found throughout the body and contains a random arrangement of collagen and elastic fibers. The strongest type is (36) _____ connective tissue, which is composed of primarily (37) _____ fibers arranged in one direction, and which is found in tendons and ligaments. The most flexible type is (38) _____, which surrounds organs like the bladder and stomach that need to change size regularly. (39) _____ connective tissue is meshlike, consisting of branched reticular fibers, and forms the internal framework of soft organs.

Special connective tissue is very diverse, and serves specific functions in the body. Like fibrous connective tissue, most special connective tissue consists of fibers and cells embedded in a ground substance. There are four types of special connective tissue:

cartilage, bone, blood, and adipose tissue. Cartilage contains (40) _____ fibers in a gel-like ground substance that is produced by (41) _____. This tissue contains no blood vessels. Bone has a (42) _____ matrix with only a few cells. (43) _____ consists of cells suspended in a fluid matrix. Adipose tissue has few fibers and almost no ground substance, but lots of cells called (44) _____. Adipose tissue is specialized for (45) _____ storage.

4.4 Muscle tissues contract to produce movement

4.5 Nervous tissue transmits impulses

Matching

____ 1. **muscle tissue**   a. cells of the nervous tissue that support neurons

____ 2. **skeletal muscle tissue**   b. muscle tissue that surrounds hollow organs

____ 3. **cardiac muscle tissue**   c. a tissue specialized for movement

____ 4. **smooth muscle tissue**   d. nervous tissue cells that generate and transmit electrical impulses

____ 5. **nervous tissue**   e. muscle tissue connected to bone by tendons

____ 6. **neurons**   f. muscle tissue found in the heart

____ 7. **glial cells**   g. a tissue specialized for generating and transmitting electrical impulses

Completion

Complete the table by filling in the characteristics of each type of muscle tissue.

Tissue Type	Location	Number of Nuclei	Voluntary/Involuntary
8. Cardiac muscle cells	a.	b.	c.
9. Skeletal muscle cells	a.	b.	c.
10. Smooth muscle cells	a.	b.	c.

4.6 Organs and organ systems perform complex functions

Fill-in-the-Blank

1. A structure composed of two or more tissue types joined together to perform a function is a(n) _____.

2. A(n) _____ system is a group of organs working together to perform a broad function.

3. A _____ membrane consists of a layer of epithelial tissue and a layer of connective tissue.

4. A general term for a thin layer that covers or surrounds something is _____.

5. _____ membranes line thin cavities between bones in movable joints.

6. Body cavities are lined by _____ membranes.

7. The skin is also referred to as a _____ membrane.

8. _____ membranes line the airways, digestive tract, and reproductive passages.

9. A _____ plane divides the body into top and bottom portions.

10. A structure nearer to a point of reference is described as _____.

11. The thoracic cavity includes the _____ cavities and the _____ cavity.

12. The thoracic and abdominal cavities constitute the _____ cavity.

13. The thoracic and abdominal cavities are separated by the _____.

Completion

Fill in the table below by naming each organ system described.

Organ System	Description
14.	Responsible for gas exchange between the air and the blood.
15.	Produces or resists movement, generates heat.
16.	Transports material through the body, helps maintain body temperature, aids body defenses.
17.	Produces eggs and sperm.
18.	Excretes wastes, maintains volume and composition of body fluids.
19.	Detects stimuli in the external and internal environment and coordinates a response.

continued

Organ System	Description
20.	Provides the structural framework for movement of the body.
21.	Provides protection from injury, infection, and dehydration; receives sensory input.
22.	Returns excess tissue fluid to the circulatory system and aids body defense mechanisms.
23.	Produces hormones that regulate many body functions.
24.	Provides the body with water and nutrients.

4.7 The skin as an organ system

Crossword Puzzle

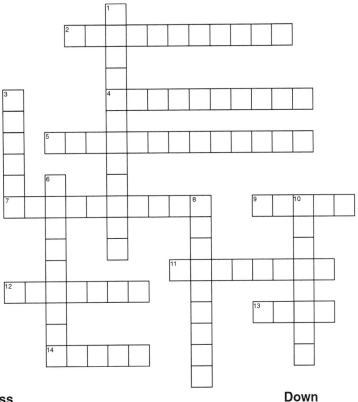

Across

2. Pigment-producing cells in the epidermis
4. Sweat glands produce sweat that contains this antibiotic secretion
5. The skin represents this organ system
7. A supporting layer for the skin that contains adipose tissue and loose connective tissue
9. Cells at the base of the epidermis that are constantly dividing
11. Small projections on the surface of the dermis
12. A waterproof protein produced by keratinocytes in the epidermis
13. The region of a hair that is found below the skin's surface
14. The region of a hair that extends above the skin's surface

Down

1. The outer layers of the epidermis are dead because these are not found in the epidermis
3. This type of muscle is found at the base of each hair follicle
6. The outer layer of the skin's epithelial tissue
8. A type of gland that produces oil
10. This type of epithelial cell is found in the epidermis

4.8 Multicellular organisms must maintain homeostasis

Ordering

Place these events of a negative feedback cycle in the correct sequence; then answer the questions that follow.

  a. core temperature returns to normal
  b. hypothalamus decreases impulses to the blood vessels and activates the sweat glands
  c. core temperature rises
  d. skin and internal organs sense a change in temperature
  e. blood vessels dilate and perspiration increases
  f. skin and internal organs send a message to the hypothalamus

1. _____

2. _____

3. _____

4. _____

5. _____

6. _____

_____ 7. What is the controlled variable?

_____ 8. What acts as the sensor?

_____ 9. What acts as the control center?

_____ 10. What acts as the effector?

_____ 11. What type of feedback will amplify a change rather than return body systems to normal?

Chapter Test

Multiple Choice

1. When a group of cells perform a common function they become a(n):
   a. organism.
   b. organ.
   c. tissue.
   d. organ system.

2. _____ tissue covers or lines body surfaces.
   a. Connective
   b. Adipose
   c. Loose
   d. Epithelial

3. A simple epithelium would *not* function in:
   a. transporting materials.
   b. absorbing nutrients.
   c. secreting wastes.
   d. providing protection.

4. All of the following are associated with exocrine glands *except:*
   a. saliva.
   b. ducts.
   c. hormones.
   d. sweat glands.

5. A basement membrane consists of _____ secreted by _____.
   a. carbohydrates, fibroblasts
   b. proteins, epithelial cells
   c. proteins, connective tissue cells
   d. fat, adipocytes

6. Matrix is found in all of the following *except:*
   a. adipose tissue.
   b. cartilage.
   c. elastic connective tissue.
   d. muscle.

7. Cardiac muscle cells must be able to communicate rapidly with each other, a process that involves the movement of ions from one cell to another. Which type of junction aids this process in cardiac muscle cells?
   a. tight junctions
   b. communication junctions
   c. adherence junctions
   d. gap junctions

8. A unique feature of connective tissue is that the cells of the tissue secrete the _____.
   a. plasma membrane
   b. junctions
   c. basement membrane
   d. matrix

9. Which of the following connective tissue elements are incorrectly matched with a characteristic?
   a. elastic fibers–allow tissue to stretch
   b. reticular fibers–secrete matrix
   c. collagen fibers–confer strength
   d. adipocytes–store fat

10. Fibroblasts are found in:
    a. epithelium.
    b. nervous tissue.
    c. muscle tissue.
    d. connective tissue.

11. Cartilage is slow to heal because:
    a. it is a connective tissue.
    b. it is not specialized for growth.
    c. it is seldom injured.
    d. it has no blood vessels.

12. The cells of muscle tissue are called:
    a. adipocytes.
    b. fibroblasts.
    c. chondroblasts.
    d. fibers.

13. The walls of blood vessels contain _____ muscle that is _____.
    a. skeletal muscle, voluntary
    b. skeletal muscle, involuntary
    c. smooth muscle, voluntary
    d. smooth muscle, involuntary

14. In a neuron the _____ receives signals, and the _____ transmits signals.
    a. axon, cell body
    b. cell body, dendrite
    c. dendrite, axon
    d. axon, dendrite

15. The posterior cavity consists of:
    a. the thoracic cavity and the spinal cavity.
    b. the cranial cavity and the thoracic cavity.
    c. the thoracic cavity and the abdominal cavity.
    d. the cranial cavity and the spinal cavity.

16. Tissue membranes consist of:
    a. epithelial tissue and connective tissue.
    b. epithelial tissue and a basement membrane.
    c. a basement membrane and a cellular membrane.
    d. a pleural membrane and a parietal membrane.

17. Goblet cells are found in:
    a. some epithelial tissue.
    b. connective tissue.
    c. mucous membranes.
    d. the hypodermis.

18. Which of the following systems produces hormones that regulate many body functions?
    a. integumentary
    b. nervous
    c. endocrine
    d. reproductive

19. A _____ plane divides the body into right and left segments.
    a. transverse
    b. midsagittal
    c. frontal
    d. proximal

20. The skin rests on a supportive layer called the:
    a. dermis.
    b. epidermis.
    c. hyperdermis.
    d. hypodermis.

21. Actively dividing keratinocytes are referred to as:
    a. dermal cells.
    b. basal cells.
    c. fibroblasts.
    d. melanocytes.

22. The epidermis lacks:
    a. pigment.
    b. keratin.
    c. melanin.
    d. blood vessels.

23. A drop in blood pressure results in constriction of the blood vessels. Blood pressure functions as the _____ in a negative feedback system.
    a. sensor
    b. effector
    c. control center
    d. controlled variable

24. As humans age the skin begins to sag and wrinkle. This is most likely due to a change in:
    a. the cellular junctions.
    b. the blood supply to the skin.
    c. the type of collagen fibers produced.
    d. the amount of skeletal muscle present.

25. Contractions of the uterus during childbirth are an example of positive feedback because:
    a. they are a short-term event.
    b. they are not required to maintain health.
    c. a change in a controlled variable is not monitored by sensors.
    d. a change in a controlled variable serves to amplify the change.

Key Concept Review Questions

Each of the Key Concepts listed at the beginning of this chapter has been rewritten as a question below. After successfully completing the study guide exercises and the Chapter Test, you should be able to answer each of these questions. Refer to the Key Concepts list at the beginning of this chapter to check your answers.

1. What are some advantages of being multicellular?
2. What is a tissue? What are the four major tissue types in the body?
3. Where are epithelial tissues found, and how are they classified?
4. What are glands? What is their function?
5. What is the function of a basement membrane? What is the function of a cellular junction?
6. What is the broad function of connective tissue?
7. What components make up connective tissue?
8. What are the two broad categories of connective tissue?
9. What are four types of fibrous connective tissue?
10. What are four types of specialized connective tissue?

11. How are connective tissue types different?
12. What is the function of muscle tissue? List three types of muscle tissue.
13. What are three characteristics of skeletal muscle cells?
14. What are three characteristics of smooth muscle cells?
15. What are three characteristics of cardiac muscle cells?
16. What is the function of nervous tissue? Name two types of nervous tissue cells and their function.
17. What is an organ system? How many organ systems are found in the human body?
18. What is the function of a tissue membrane? Name the four major types of tissue membranes.
19. What are the two main components of the integumentary system? What are the functions of this system?
20. How do multicellular organisms maintain homeostasis? What is involved in this mechanism?

5

The Skeletal System

Chapter Summary and Key Concepts

After reading and studying this chapter you should know the following:

Sections 5.1, 5.2, 5.3

1. The skeletal system consists of bones, ligaments, and cartilage.

2. The skeleton provides support, movement, and protection. In addition, the bones provide for formation of blood cells and mineral storage.

3. Compact bone tissue consists of osteons containing osteocytes in lacunae that communicate via canaliculi.

4. Spongy bone tissue is a latticework of trabeculae containing osteocytes in lacunae; the spaces between the trabeculae are filled with red marrow.

5. Ligaments attach bone to bone, and tendons attach muscle to bone.

6. The skeletal system contains three types of cartilage: fibrocartilage, hyaline cartilage, and elastic cartilage.

7. Ossification is the process of developing bone from cartilage, which begins in the embryo.

8. Bones lengthen through adolescence, and undergo replacement, repair, and remodeling throughout adulthood.

9. Bone cells include osteocytes that maintain bone tissue, osteoblasts that build bone tissue, and osteoclasts that dissolve bone tissue.

10. Homeostasis of bone tissue requires a balance in the activity of osteoblasts and osteoclasts.

11. Osteoblast and osteoclast activity is regulated by parathyroid hormone and calcitonin.

Sections 5.4, 5.5, 5.6

12. Bones are classified by shape as long, short, flat, and irregular.

13. Bones consist of both compact and spongy bone tissue arranged to meet the functional needs of the bone.

14. The skeleton is organized into the axial skeleton and the appendicular skeleton.

15. The axial skeleton consists of the skull, vertebral column, ribs, and sternum.

16. The appendicular skeleton consists of the pectoral girdle, pelvic girdle, and limbs.

17. Joints are points of contact between bones and are called articulations.

18. Joints are classified in three broad categories based on mobility: fibrous joints are immovable, cartilaginous joints are slightly movable, and synovial joints are freely movable.

19. Synovial joints contain a joint cavity lined with synovial membrane and filled with synovial fluid.

20. Diseases and disorders of the skeletal system include sprains, bursitis, tendinitis, osteoarthritis, and rheumatoid arthritis.

Exercises

Complete the exercises for each section after you have read and studied the section. If you cannot answer some questions, or answer them incorrectly, return to the chapter and review this information. You may find it helpful to work on only one section at a time. When you have completed all sections, take the Chapter Test as an indicator of your mastery of this topic.

5.1 The skeletal system consists of connective tissue

5.2 Bone development begins in the embryo

5.3 Mature bone undergoes remodeling and repair

Matching

____ 1. **bone**   a. dense fibrous connective tissue structures that attach bone to bone

____ 2. **compact bone**   b. a strip of cartilage in the epiphysis of a long bone that allows for bone lengthening

____ 3. **spongy bone**   c. the hard elements of the skeleton

____ 4. **osteocytes**   d. a bone condition in which large amounts of bone mass are lost

____ 5. **osteons**   e. cartilage-forming cells

____ 6. **central canal**   f. dense bone tissue that forms the shaft of a long bone

____ 7. **ligaments**   g. bone-dissolving cells

____ 8. **cartilage**   h. a ringlike arrangement of bone tissue formed in compact bone

____ 9. **chondroblasts** i. less dense bone tissue that is light but strong

____ 10. **osteoblasts** j. a passageway through the center of an osteon that carries blood vessels

____ 11. **growth plate** k. a connective tissue that contributes to the function of the skeletal system by forming embryonic structures and aiding movement

____ 12. **osteoclast** l. bone-forming cells

____ 13. **osteoporosis** m. mature bone cells that maintain the structure of bone

Fill-in-the-Blank

Referenced sections are in parentheses.

14. The three functions of the skeleton are _____, _____, and _____. (5.1)

15. Two additional functions of bone are _____ and _____. (5.1)

16. Cartilage contains fibers of _____ and/or _____ in a ground substance of _____. (5.1)

17. The intervertebral discs are composed of _____. (5.1)

18. _____ cartilage covers the ends of mature bones. (5.1)

19. The most flexible type of cartilage is _____. (5.1)

20. Bone remodeling and repair is due to the activity of _____ and _____. (5.3)

21. Electrical currents within bone are caused by _____ and increase the activity of _____. (5.3)

22. A mass of clotted blood produced when a bone fractures is a _____. (5.3)

23. A fibrocartilage bond between two broken ends of a bone is a _____. (5.3)

Labeling

Identify each structure indicated in Figure 5.1.

Figure 5.1

24. _____   30. _____

25. _____   31. _____

26. _____   32. _____

27. _____   33. _____

28. _____   34. _____

29. _____   35. _____

Paragraph Completion

Use the following terms to complete the paragraph, and then repeat this exercise with the terms covered.

bone   ossification   hyaline   osteoblasts

lacunae   osteocytes   chondroblasts   hydroxyapatite

osteoid   matrix

In the embryo, bone development begins in a bone model made of (36) _____ cartilage. In this process, called (37) _____, the cartilage begins to dissolve and is replaced by (38) _____. This begins when (39) _____ die and their (40) _____ breaks down. Blood vessels develop in the area and bring in bone-forming (41) _____ that secrete (42) _____ and an enzyme that promotes the crystallization of hard mineral salts called (43) _____. As the bone matrix hardens, osteoblasts become trapped and develop into mature (44) _____ embedded in individual (45) _____.

Word Choice

Circle the correct term to complete each sentence.

46. Weight-bearing exercise (increases/decreases) bone mass.
47. Mature compact bone is (one-third/two-thirds) osteoid and (one-third/two-thirds) hydroxyapatite.
48. Growth hormone (stimulates/inhibits) bone lengthening at the growth plate.
49. Falling levels of calcium in the blood (increase/decrease) the activity of parathyroid hormone.
50. Rising levels of calcium in the blood (increase/decrease) the activity of calcitonin.
51. Calcitonin (increases/decreases) the activity of osteoblasts.
52. Increased osteoclast activity (increases/decreases) the blood level of calcium.

5.4 The skeleton protects, supports, and permits movement

Labeling

Identify each of the bones indicated in Figure 5.2.

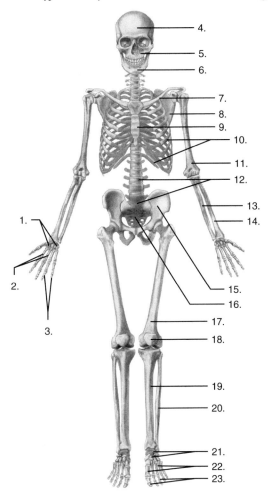

Figure 5.2

1. _____
2. _____
3. _____
4. _____
5. _____
6. _____
7. _____
8. _____
9. _____
10. _____
11. _____
12. _____
13. _____
14. _____
15. _____
16. _____
17. _____
18. _____
19. _____
20. _____
21. _____
22. _____
23. _____

True/False

_____ 24. Sinuses are air spaces found in the coxal bones.

_____ 25. The skeleton consists of 206 bones.

_____ 26. The vertebral column is classified in seven anatomical regions.

_____ 27. Herniated discs occur most often in the lumbar vertebrae.

_____ 28. The appendicular skeleton contains the pectoral girdle, the pelvic girdle, and the limbs.

5.5 Joints form connections between bones

Matching

Match each characteristic (1–10) listed below with a type of joint (a, b, or c).

a. synovial joint   **b. fibrous joint**   **c. cartilaginous joint**

____ 1. strengthened by ligaments and tendons

____ 2. found between cranial bones

____ 3. a freely movable joint

____ 4. connects the vertebrae of the backbone

____ 5. may contain menisci to reduce friction

____ 6. includes hinge joints and ball-and-socket joints

____ 7. an immovable joint

____ 8. has a joint capsule

____ 9. may include bursa for cushioning

____ 10. a slightly movable joint

5.6 Diseases and disorders of the skeletal system

Completion

Name each injury or disorder of the skeletal system described in the table.

Injury or disorder of the skeletal system	Description
1.	joint inflammation caused by a malfunction of the immune system
2.	stretched or torn ligaments accompanied by bruising, swelling, and pain
3.	inflammation of the bursae following injury
4.	joint inflammation resulting from degenerative "wear and tear"

Chapter Test

Multiple Choice

1. Which of the following is not a function of the skeleton?
   a. support
   b. formation of blood cells
   c. regulation of blood cell production
   d. mineral storage

2. The shaft of a mature long bone is composed of _____ and filled with _____.
   a. compact bone, red marrow
   b. compact bone, yellow marrow
   c. spongy bone, red marrow
   d. spongy bone, yellow marrow

3. The periosteum:
   a. lines the inner cavity of a long bone and contains the bone marrow.
   b. covers the ends of a long bone to reduce friction in the joint.
   c. is found at the junction of compact bone and spongy bone.
   d. covers the outer surface of the bone and contains osteoblasts.

4. Osteocytes in compact bone remain in contact with each other via cytoplasmic extensions that extend through:
   a. lacunae.
   b. trabeculae.
   c. canaliculi.
   d. a central canal.

5. Spongy bone and compact bone tissue both consist of a hard mineralized matrix containing calcium and phosphate.
   a. true
   b. false

6. A structure composed of dense fibrous connective tissue that attaches bone to bone is a:
   a. ligament.
   b. tendon.
   c. osteon.
   d. trabeculae.

7. Hyaline cartilage is found in all of the following *except:*
   a. embryonic structures that will become bone.
   b. the ends of long bones.
   c. cartilaginous joints.
   d. tendons and ligaments.

8. Which of the following occurs first in the process of ossification?
   a. Blood vessels develop in the cartilage model.
   b. Osteoblasts begin building bone tissue.
   c. Chondroblasts die.
   d. Growth plates allow lengthening.

9. A mature bone that had a lower-than-normal percentage of hydroxyapatite would be:
   a. too flexible.
   b. too brittle.
   c. too long.
   d. too short.

10. Bones can no longer lengthen when:
    a. the periosteum stops developing.
    b. the cartilage growth plate is replaced with bone.
    c. the number of osteoblasts in the growth plate increases.
    d. the spongy bone is converted to compact bone.

11. Which of the following is incorrectly matched?
    a. osteoclasts—dissolve bone tissue
    b. osteoblasts—bone-forming cells
    c. osteocytes—bone remodeling and repair cells
    d. chondroblast—cartilage-forming cells

12. The bones of trained athletes may be thicker and heavier than the bones of nonathletes because:
    a. athletes generally have better nutrition.
    b. athletic activity stimulates higher hormone levels.
    c. bone mass increases when bone is stressed.
    d. athletic activity decreases the activity of osteocytes.

13. Parathyroid hormone stimulates the activity of:
    a. osteocytes.
    b. osteoblasts.
    c. osteoclasts.
    d. chondroblasts.

14. Adequate calcium intake is important in the prevention of osteoporosis. Which of the following might occur if calcium intake was too low?
    a. Low blood calcium levels would stimulate the activity of osteoblasts.
    b. Low blood calcium levels would stimulate the activity of osteoclasts.
    c. Inadequate calcium would be available for deposit into bone tissue.
    d. Osteocytes would die.
    e. b and c

15. The cranial bones and sternum are examples of _____ bones.
    a. long
    b. short
    c. flat
    d. irregular

16. Fused vertebrae are found in the _____ region of the vertebral column.
    a. lumbar
    b. coccygeal
    c. thoracic
    d. cervical

17. Cartilage discs between the vertebrae are composed of _____ and function _____.
    a. fibrocartilage, as shock absorbers
    b. hyaline cartilage, as protection
    c. hyaline cartilage, as attachments between vertebrae
    d. elastic cartilage, in flexibility

18. Floating ribs:
    a. do not attach to the vertebral column.
    b. do not attach to the sternum.
    c. do not attach to any muscles.
    d. do not attach to the rib cage.

19. The fingers are composed of:
    a. carpals.
    b. phalanges.
    c. metacarpals.
    d. tarsals.

20. The disadvantage of highly flexible joints is:
    a. inadequate blood supply.
    b. poor control of movement.
    c. poor stability.
    d. limited movement.

21. Which of the following correctly sequences the joints in order of increasing mobility?
    a. synovial, fibrous, cartilaginous
    b. fibrous, synovial, cartilaginous
    c. cartilaginous, fibrous, synovial
    d. fibrous, cartilaginous, synovial

22. The knee joint contains all of the following *except:*
    a. menisci.
    b. bursa.
    c. synovial fluid.
    d. fontanels.

23. A movement that increases the angle of a joint is called:
    a. supination.
    b. rotation.
    c. extension.
    d. circumduction.

24. Inflammation of the joints due to a malfunction of the immune system is called:
    a. osteoarthritis.
    b. immunoarthritis.
    c. rheumatoid arthritis.
    d. inflammatory arthritis.

25. Which of the following does not apply to sprains?
    a. rapid healing due to a rich blood supply
    b. stretched or torn ligaments
    c. internal bleeding
    d. possible surgery to remove a torn ligament

Key Concept Review Questions

Each of the Key Concepts listed at the beginning of this chapter has been rewritten as a question. After successfully completing the study guide exercises and the Chapter Test, you should be able to answer each of these questions. Refer to the Key Concepts list at the beginning of this chapter to check your answers.

1. What three types of connective tissue are found in the skeletal system?
2. What are three functions of the skeleton and two additional functions of bone?
3. How would you describe the structure of compact bone?
4. How would you describe the structure of spongy bone?
5. What is the function of a ligament? What is the function of a tendon?
6. What three types of cartilage are found in the skeletal system?
7. What is ossification? When does it begin?
8. What bone activities occur in adolescence and in adulthood?
9. What are the names of three types of bone cells and the function of each?
10. What is required for homeostasis of bone tissue?
11. What two hormones regulate the activity of osteoblasts and osteoclasts?

12. What are the four categories of shape used to classify bones?
13. What type of bone tissue is found in each shape of bone?
14. What are the two divisions of the skeleton?
15. What is included in the axial skeleton?
16. What is included in the appendicular skeleton?
17. What is a joint? What is another term for a joint?
18. What are the names of the three categories of joints and the mobility of each?
19. How would you describe the structure of a synovial joint?
20. What are the five disorders of the skeletal system discussed in this chapter?

6

The Muscular System

Chapter Summary and Key Concepts

After reading and studying this chapter you should know the following:

Sections 6.1, 6.2

1. Muscles produce movement, generate tension, and generate heat.

2. The three types of muscle are skeletal muscle, cardiac muscle, and smooth muscle.

3. Muscle cells are excitable and act by contracting and then relaxing.

4. Muscles most frequently attach to bone via tendons, but they may also attach to other muscles or to the skin.

5. Muscles have one attachment point called the origin and a second attachment point called the insertion. When muscles contract, the insertion is pulled toward the origin.

6. In skeletal muscle, individual muscle cells are called fibers, and they are arranged in bundles called fascicles. A whole muscle contains many fascicles bound together with connective tissue. The connective tissue layers of a muscle extend beyond the muscle as a tendon.

7. Muscle fibers contain myofibrils, which in turn contain the contractile filaments actin and myosin.

8. The contractile unit of a muscle is the sarcomere.

9. Skeletal muscle cells are stimulated by motor neurons that release the neurotransmitter acetylcholine.

10. During contraction, acetylcholine stimulates the release of calcium ions into the cytoplasm. The presence of calcium triggers the events leading to contraction.

11. Contraction results when actin filaments are pulled across myosin filaments, causing the sarcomere to shorten.

12. Muscle contraction and relaxation require ATP.

Section 6.3

13. The force of muscle contraction is determined by the number of muscle cells in a motor unit, the number of motor units involved, and the frequency of nerve stimulation.

14. A twitch is a complete cycle of contraction and relaxation. A latent period precedes a twitch.

15. Recruitment is an increase in muscle force due to an increasing number of motor units contracting.

16. Summation is an increase in muscle force due to an increase in the frequency of nerve stimulation.

17. The internal structure and chemistry of a muscle fiber will determine whether it is a slow-twitch fiber or a fast-twitch fiber. Muscle strength and endurance are determined by the number of slow-twitch and fast-twitch fibers present.

Sections 6.4, 6.5

18. Cardiac muscle cells have one nucleus, intercalated discs, and striations; are involuntary; and are found only in the heart.

19. Smooth muscle cells have one nucleus, gap junctions, and no striations; are involuntary; and are found around hollow organs or tubes.

20. Disorders of the muscular system include muscular dystrophy, tetanus, muscle cramps, pulled muscles, and fasciitis.

Exercises

Complete the exercises for each section after you have read and studied the section. If you cannot answer some questions, or answer them incorrectly, return to the chapter and review this information. You may find it helpful to work on only one section at a time. When you have completed all sections, take the Chapter Test as an indicator of your mastery of this topic.

6.1 Muscles produce movement or generate tension

6.2 Individual muscle cells contract and relax

Matching

_____ 1. **skeletal muscle** a. the contractile unit of a muscle cell

_____ 2. **origin** b. a process resulting in shortening of a sarcomere

_____ 3. **insertion** c. a decline in muscle performance during sustained exercise

_____ 4. **myofibrils** d. muscles that attach to bone

_____ 5. **sarcomere**	e. a specialized smooth endoplasmic reticulum that stores calcium ions in muscle cells

_____ 6. **myosin**	f. point of attachment of a skeletal muscle to a structure that remains stationary

_____ 7. **actin**	g. a condition that occurs when muscles use more ATP than can be supplied by aerobic respiration

_____ 8. **neuromuscular junction**	h. cylindrical structures within a muscle cell that are packed with actin and myosin

_____ 9. **sarcoplasmic reticulum**	i. the protein found in thick filaments

_____ 10. **sliding filament mechanism**	j. point of attachment of a skeletal muscle to a structure that moves

_____ 11. **oxygen debt**	k. the junction between a motor neuron and a skeletal muscle cell

_____ 12. **fatigue**	l. the protein found in thin filaments

Fill-in-the-Blank

Referenced sections are in parentheses.

13. Muscles can _____ or _____ movement or _____ tension. (6.1)

14. When muscle cells respond to chemical and/or electrical signals, they are referred to as _____. (6.1)

15. When muscle cells _____, they shorten; when they _____, they return to their original length. (6.1)

16. Muscle groups that work together to create the same movement are called _____. (6.1)

17. Muscles that oppose each other are called _____. (6.1)

18. When a muscle contracts, the _____ is pulled toward the _____. (6.1)

19. The segment of a myofibril from one Z line to the next Z line is a _____. (6.1)

20. When a muscle is relaxed, cross-bridge formation is prevented by the position of the _____ _____. (6.2)

Labeling

Label each structure indicated in Figure 6.1.

Figure 6.1

Paragraph Completion

Use the following terms to complete the paragraph, and then repeat this exercise with the terms covered. Terms used more than once are listed multiple times.

electrical	sarcoplasmic reticulum	ATP
actin		neurotransmitter
troponin	contraction	cross-bridges
T tubules	lengthens	sarcoplasmic reticulum
bend	relaxation	
ATP	motor neuron	troponin-tropomyosin complex
sliding-filament	cross-bridges	
acetylcholine	binding	shortening

The events leading to muscle contraction begin at the neuromuscular junction with the release of (33) _____ from a (34) _____. Acetylcholine is a(n) (35) _____ that attaches to (36) _____ sites on the muscle cell. This causes the muscle cell to generate a(n) (37) _____ impulse that is carried deep into the cell by (38) _____. When the impulse reaches the (39) _____, calcium ions are released into the cytoplasm. The calcium ions bind to the (40) _____ molecule, causing a shift in the position of the (41) _____ and exposing myosin binding sites. With the binding sites exposed, myosin heads form (42) _____ with the actin, then (43) _____ and pull the (44) _____ filament toward the center of the sarcomere. This is called the (45) _____ mechanism of contraction and results in (46) _____ of the sarcomere and (47) _____ of the muscle. (48) _____ is required as an energy source. (49) _____ occurs when nerve stimulation ends. During relaxation, calcium ions are transported back into the (50) _____, and the troponin-tropomyosin complex again blocks the myosin binding sites, preventing the formation of (51) _____. The filaments slide back to their original position and the muscle cell (52) _____. (53) _____ is also required for relaxation.

Completion

Answer the following questions about Figure 6.2.

_____ 54. What represents myosin?

_____ 55. What is associated with the troponin-tropomyosin complex?

_____ 56. What has heads that form cross-bridges?

_____ 57. What slides toward the center of the sarcomere during a contraction?

_____ 58. What has myosin binding sites?

Figure 6.2

6.3 The activity of muscles can vary

Matching

___ 1. **motor unit** a. a complete cycle of contraction and relaxation

___ 2. **muscle tension** b. increasing muscle force by increasing the rate of stimulation of motor units

___ 3. **all-or-none principle** c. a motor neuron and all the muscle cells it stimulates

___ 4. **twitch** d. the mechanical force that muscles generate when they contract

___ 5. **recruitment** e. muscle cells respond fully every time they are stimulated

___ 6. **summation** f. increasing muscle force by activating more motor units

Fill In

Indicate which of the following characteristics apply to each type of muscle fiber.

    a. contract quickly

    b. contract slowly

    c. have an extensive supply of blood vessels

    d. have fewer blood vessels

    e. rely primarily on aerobic metabolism

    f. rely on stored glycogen and creatine phosphate

    g. fatigue quickly

    h. can sustain a contraction without fatiguing

    i. contain many mitochondria

    j. contain fewer mitochondria

    k. are well adapted for endurance sports

    l. are best for sports requiring quick bursts of energy

    m. have little or no myoglobin

    n. store large amounts of myoglobin

7. Slow-twitch fibers:

8. Fast-twitch fibers:

Short Answer

9. Why does summation increase muscle force during a contraction?

10. What causes tetanus?

11. What are the two primary types of exercise training?

12. Why does strength training have a greater effect on fast-twitch fibers than slow-twitch fibers?

13. Why do most marathon runners have a lean build, while weight lifters exhibit greater muscle mass?

14. A friend has just joined the gym and excitedly explains to you that he is taking amino acid supplements to help him make more muscle cells, which he hopes will result in an increase in muscle mass. How would you respond?

6.4 Cardiac and smooth muscles have special features

6.5 Diseases and disorders of the muscular system

Matching

List each muscle type beside all applicable characteristics. Muscle types may be used more than once, and several muscle types may correspond with a characteristic.

a. **skeletal muscle** b. **cardiac muscle** c. **smooth muscle**

_____ 1. voluntary muscle

_____ 2. involuntary muscle

_____ 3. have intercalated discs

_____ 4. striated

_____ 5. non-striated

_____ 6. joined by gap junctions

_____ 7. attached to bones

_____ 8. can only contract when stimulated by a nerve cell

Short Answer

For each muscular disorder listed below, indicate the cause and the symptoms.

9. Muscular Dystrophy:
   a. Cause:
   b. Symptoms:

10. Tetanus:
    a. Cause:
    b. Symptoms:

11. Muscle Cramps:
    a. Cause:
    b. Symptoms:

12. Pulled Muscle:
    a. Cause:
    b. Symptoms:

13. Fasciitis:
    a. Cause:
    b. Symptoms:

Chapter Test

Multiple Choice

1. Muscle movements over which we have conscious control are:
   a. voluntary.
   b. involuntary.
   c. specific.
   d. natural.

2. The biceps brachii, the brachialis, and brachioradialis all contribute to some of the same movements. These muscles would be:
   a. antagonistic.
   b. synergistic.
   c. complementary.
   d. cooperative.

3. When a skeletal muscle contracts, the _____ is pulled toward the _____.
   a. origin, insertion
   b. origin, muscle
   c. insertion, origin
   d. insertion, muscle

4. Arrange the following components of a muscle in order of increasing size:
   a. fiber, myofibril, fascicle, muscle.
   b. fascicle, myofibril, fiber, muscle.
   c. muscle, myofibril, fascicle, fiber.
   d. myofibril, fiber, fascicle, muscle.

5. The contractile unit of a muscle is a(n):
   a. myofibril.
   b. fascicle.
   c. sarcomere.
   d. fiber.

6. In the sliding-filament mechanism of muscle contraction, _____ slides over _____.
   a. Z lines, actin
   b. actin, myosin
   c. myosin, actin
   d. Z lines and actin, myosin

7. Which of the following would interfere most with a muscle's ability to contract?
   a. absence of acetylcholine binding sites on the muscle membrane
   b. a reduced number of actin and myosin filaments
   c. an increased level of intracellular calcium
   d. a reduced number of T tubules

8. In a muscle cell _____ contain _____.
   a. myofibrils, actin and myosin filaments
   b. actin and myosin filaments, myofibrils
   c. sarcomeres, myofibrils
   d. myofibrils, sarcomeres
   e. a and d

9. Which of the following best describes the action of sarcomeres within a muscle cell during a contraction?
    a. Each sarcomere shortens a little.
    b. Only the sarcomeres on the ends of the muscle shorten in a weak contraction.
    c. Sarcomeres in the center of the muscle shorten more than sarcomeres on the ends of the muscle.
    d. Sarcomeres shorten in different degrees depending on the amount of actin and myosin they contain.

10. The force of a muscle contraction can be increased by:
    a. increasing the frequency of nerve stimulation.
    b. decreasing the number of motor units responding.
    c. increasing the amount of collagen in the tendons.
    d. decreasing the amount of Ca^{++} released from the sarcoplasmic reticulum.

11. Place the following events of muscle contraction in the proper sequence.
    I. calcium is released from the sarcoplasmic reticulum
    II. cross-bridges form
    III. actin slides across myosin
    IV. calcium binds to troponin, causing the troponin-tropomyosin complex to shift its position

    a. I, II, IV, III
    b. I, IV, II, III
    c. II, I, IV, III
    d. III, I, IV, II

12. In the formation of a cross-bridge, _____ binds to _____.
    a. myosin, actin
    b. actin, troponin
    c. myosin, tropomyosin
    d. tropomyosin, actin

13. When a muscle relaxes, ATP is required. The reason for this is:
    a. the actin filaments must be pushed back to their original position.
    b. the myosin heads can only detach from the actin when a molecule of ATP binds to the head.
    c. energy is needed to pull the sarcomeres apart.
    d. the connective tissue elements of the muscle require energy to resume their original position.

14. Which of the following represents the correct sequence in which a muscle cell will access energy sources?
    a. creatine phosphate, stored ATP, glycogen, aerobic metabolism
    b. aerobic metabolism, creatine phosphate, glycogen, stored ATP
    c. stored ATP, glycogen, creatine phosphate, aerobic metabolism
    d. stored ATP, creatine phosphate, glycogen, aerobic metabolism

15. A muscle contraction that results in movement is called:
    a. isometric.
    b. isotonic.
    c. tetanic.
    d. metabolic.

16. The physiological cause of muscle fatigue is:
    a. inadequate sarcomeres to meet muscle demands.
    b. inadequate ATP to meet metabolic demands.
    c. inadequate numbers of slow-twitch fibers.
    d. constriction of blood vessels as lactic acid accumulates.

17. A motor unit consists of:
    a. one motor neuron and all the muscle cells it stimulates.
    b. one muscle and all the motor neurons that stimulate the muscle.
    c. one motor neuron and one muscle cell.
    d. all the motor neurons and all the muscle cells of one whole muscle.

18. The all-or-none principle states that:
    a. when a muscle is stimulated, every motor unit will respond.
    b. when a muscle cell is stimulated, it responds with a complete cycle of contraction and relaxation.
    c. when a muscle cell is stimulated, the sarcomeres that respond will shorten.
    d. a small stimulus can produce a small contraction.

19. Increasing muscle force by increasing the number of motor units responding is called:
    a. summation.
    b. recruitment.
    c. tetany.
    d. magnification.

20. When a muscle cell is stimulated too frequently to allow relaxation, the muscle will:
    a. fail to respond.
    b. tear and bleed.
    c. have tetanic contractions.
    d. contract with more force than normally possible when the maximum number of motor units contract.

21. Which of the following would *not* be found in a slow-twitch fiber?
    a. large amounts of myoglobin
    b. an extensive network of blood vessels
    c. few mitochondria
    d. reddish color

22. Strength training involves:
    a. improving muscle endurance.
    b. providing resistance that requires muscles to work harder.
    c. increasing the body's ability to supply oxygen to the muscles.
    d. increasing the number of muscle cells.

23. Cardiac muscle and smooth muscle are similar in all of the following ways *except:*
    a. both have muscle cells joined by gap junctions.
    b. the contraction of both can be modified by nerve stimulation.
    c. both are involuntary muscles.
    d. both fatigue easily.

24. Inflammation of the fascia around a muscle is called:
    a. fasciitis.
    b. tendonitis.
    c. bursitis.
    d. myositis.

25. Internal bleeding, swelling, and pain occur with:
    a. tetanus.
    b. muscle cramps.
    c. pulled muscles.
    d. fasciitis.

Key Concept Review Questions

Each of the Key Concepts listed at the beginning of this chapter has been rewritten as a question below. After successfully completing the study guide exercises and the Chapter Test, you should be able to answer each of these questions. Refer to the Key Concepts list at the beginning of this chapter to check your answers.

1. What are three things muscles can do?
2. What are the three types of muscle tissue?
3. How do muscle cells respond when they are stimulated?
4. What structures can be attachment points for muscles?
5. What are the origin and insertion? How do their positions change when a muscle contracts?
6. How would you describe the anatomy of a skeletal muscle?
7. Where are myofibrils located? What do myofibrils contain?
8. What is the contractile unit of a muscle cell?
9. What is the role of a motor neuron in stimulating skeletal muscle cells?
10. What is the effect of acetylcholine on a muscle cell?
11. What causes a sarcomere to shorten? What happens to a muscle cell when the sarcomere shortens?
12. Is ATP required for muscle contraction and relaxation?
13. What affects the force of a muscle contraction?
14. What is a twitch? What precedes a twitch?
15. What is recruitment?
16. What is summation?
17. What characterizes a muscle fiber as slow-twitch or fast-twitch? What qualities of the muscle are affected by the number of slow-twitch and fast-twitch fibers?
18. What are the characteristics of cardiac muscle?
19. What are the characteristics of smooth muscle?
20. What are five disorders of the muscular system described in this chapter?

7

Blood

Chapter Summary and Key Concepts

After reading and studying this chapter, you should know the following:

Section 7.1

1. The functions of blood include transport, regulation, and defense.

2. Blood is composed of plasma and formed elements.

3. Plasma is mostly water and also contains ions, proteins, hormones, gases, nutrients, and wastes.

4. Important plasma proteins include albumins that maintain water balance, globulins that transport substances and function as antibodies, and clotting proteins.

5. Formed elements include red blood cells called erythrocytes, white blood cells called leukocytes, and platelets.

6. All formed elements are produced by stem cells in the bone marrow.

7. Mature red blood cells have no nucleus, and they contain an oxygen-binding protein called hemoglobin that transports oxygen from the lungs to the body, and some carbon dioxide from the body to the lungs.

8. The liver and spleen destroy old and damaged red blood cells, recycling the iron atoms and converting heme groups to bilirubin.

9. Red blood cell production is controlled by the hormone erythropoietin.

10. White blood cells defend the body. They are classified as granular or agranular depending on their staining properties.

11. Granular leukocytes include neutrophils, eosinophils, and basophils. Agranular leukocytes include moncytes and lymphocytes.

12. Platelets are essential to the process of blood clotting. Their production is regulated by the hormone thrombopoietin.

Section 7.2

13. Hemostasis is the process of stopping blood loss, which occurs through vascular spasm, formation of a platelet plug, and coagulation.

14. Formation of a blood clot requires clotting factors, including prothrombin activator, thrombin, and fibrinogen.

Section 7.3

15. Blood type determines blood compatibility for transfusions.

16. Blood type is determined by cell surface proteins called antigens on the red blood cells.

17. Two common blood typing systems are the ABO system and the Rh system.

18. Transfusion reactions occur when foreign antigens are attacked by antibodies.

Section 7.4

19. Blood disorders may involve problems with any of the components of the blood.

20. Important blood disorders include carbon monoxide poisoning, anemia, leukemia, multiple myeloma, mononucleosis, blood poisoning, and thrombocytopenia.

Exercises

Complete the exercises for each section after you have read and studied the section. If you cannot answer some questions, or answer them incorrectly, return to the chapter and review this information. You may find it helpful to work on only one section at a time. When you have completed all sections, take the Chapter Test as an indicator of your mastery of this topic.

7.1 The components and functions of blood

Matching

___ 1. **formed elements**
___ 2. **plasma**
___ 3. **plasma proteins**
___ 4. **albumins**
___ 5. **globulins**
___ 6. **lipoprotein**
___ 7. **clotting proteins**
___ 8. **erythrocytes**
___ 9. **hemoglobin**
___ 10. **oxyhemoglobin**
___ 11. **deoxyhemoglobin**
___ 12. **hematocrit**
___ 13. **polycythemia**
___ 14. **stem cells**
___ 15. **erythroblasts**
___ 16. **macrophages**
___ 17. **bilirubin**
___ 18. **jaundice**
___ 19. **granulocytes**
___ 20. **agranulocytes**

a. red blood cells
b. hemoglobin with four oxygen molecules attached
c. a disorder of the bone marrow characterized by an unusually large number of red blood cells
d. cells in the red bone marrow that divide to produce all blood cells and platelets
e. the liquid portion of the blood
f. plasma proteins that transport substances in the blood
g. hemoglobin that has given up its oxygen
h. large white blood cells that engulf and destroy debris and invaders by phagocytosis
i. the largest group of solutes in the plasma
j. a yellowish pigment produced by the breakdown of heme
k. plasma proteins that help maintain water balance in the body
l. a condition in which the skin and mucous membranes appear yellow
m. immature cells that give rise to erythrocytes
n. the category of leukocytes containing monocytes and lymphocytes
o. the cellular component of blood
p. an oxygen-binding protein found in red blood cells
q. white blood cells containing granules that are more visible after staining
r. a plasma protein attached to a fat molecule
s. the percentage of the blood that consists of red blood cells
t. plasma proteins important in blood clotting

Fill-in-the-Blank

Referenced sections are indicated in parentheses.

21. Formed elements of the blood include _____ _____ _____, _____ _____ _____, and _____.(7.1)

22. Red blood cells function primarily as carriers for
    _____ and _____. (7.1)

23. Mature red blood cells lack a _____ and most
    _____. (7.1)

24. Hemoglobin contains a heme group with a(n) _____
    at the center. (7.1)

25. The life span of a red blood cell is approximately
    _____. (7.1)

26. Worn-out red blood cells are destroyed by the
    _____ and the _____. (7.1)

27. White blood cells function in _____ the body. (7.1)

28. Platelets are formed by large cells called _____.
    (7.1)

29. Platelet production is regulated by the hormone
    _____. (7.1)

30. The most abundant type of white blood cells are
    _____. (7.1)

31. The category of agranulocytes that defend the body against large
    parasites is _____. (7.1)

32. The inflammatory response is initiated by release of
    _____ from _____. (7.1)

33. Large white blood cells that differentiate into macrophages are
    _____. (7.1)

34. _____ are agranulocytes important to the
    immune system and are classified as B cells or T cells. (7.1)

35. Neutrophils seldom survive longer than _____.
    (7.1)

36. Dead and injured white blood cells are removed from the blood by
    the _____ and the _____. (7.1)

37. Platelets are produced by _____, and this produc-
    tion is regulated by the hormone _____. (7.1)

Word Choice

Circle the accurate term or phrase to complete each sentence.

38. Mature red blood cells have (no/multiple) nuclei.
39. Hemoglobin is a (protein/blood cell).
40. Hemoglobin binds O_2 best when the concentration of O_2 is (high/low).
41. Hemoglobin releases O_2 into the body tissues when the pH of the body tissues are (lower than/higher than) the pH of the blood.
42. Hemoglobin releases O_2 into the body tissues at a faster rate when body heat (increases/decreases).
43. CO_2 and O_2 bind to hemoglobin at (the same site/different sites).
44. The percentage of the blood that consists of (white blood cells/red blood cells) is called the hematocrit.
45. Most white blood cells have a (shorter/longer) life span than a red blood cell.
46. (White blood cells/Red blood cells) remain in the blood, while (white blood cells/red blood cells) can leave the blood and circulate in the tissue fluids.
47. (White blood cells/Red blood cells) are part of the body's defense system.

Completion

Fill in the requested information for each type of white blood cell listed in the table.

Mature White Blood Cell	What type of stem cell gives rise to this cell?	Is this cell granular or agranular?	What function will this cell serve?
Neutrophil	48.	49.	50.
Eosinophil	51.	52.	53.
Basophil	54.	55.	56.
Monocyte	57.	58.	59.
Lymphocyte	60.	61.	62.

Fill-in-the-Blank

Complete the following flowchart representing negative feedback control of O_2 availability.

low O_2 availability in the blood
↓

O_2 sensitive cells in the (47)_____ increase production of (48)_____

↓

(49)_____ stimulates increased production of (50)_____ by stem cells in the (51)_____

↓

increased number of (52)_____ returns O_2 availability to normal

↓

normal O_2 availability in the blood

7.2 Hemostasis: Stopping blood loss

Paragraph Completion

Use the following terms to complete the paragraph, and then repeat this exercise with the terms covered.

prothrombin	hemostasis	platelet plug
coagulation	prothrombin activator	fibrin
hemophilia	vessel	clot
vascular spasm	clotting factors	thrombin
fibrinogen	contract	platelets

The natural process of stopping the loss of blood from ruptured vessels is called (1)_____ and occurs in three stages. The first stage is (2)_____, during which blood vessels (3)_____ to reduce blood flow. The second stage involves formation of a (4)_____, in which platelets stick to other (5)_____ and to (6)_____ walls, sealing the injured area. The third stage is (7)_____. Substances known as (8)_____ participate in the formation of a blood clot. Damage to vessels

continued

stimulates the release of (9)_____, which converts (10)_____ to the enzyme (11)_____. This enzyme in turn converts the plasma protein (12)_____ into long, soluble protein threads called (13)_____. These threads interact with platelets and trapped red blood cells to form a (14)_____. Individuals with (15)_____, an inherited condition, lack one or more clotting factors and are unable to form effective blood clots.

7.3 Blood types determine blood compatibility

Fill-in-the-Blank

Referenced sections are indicated in parentheses.

1. The success of a blood transfusion depends largely on _____. (7.3)

2. Cells have surface _____ that are recognized by the immune system as "self." (7.3)

3. Foreign cells carry surface proteins called _____ that are recognized by the immune system as "nonself." (7.3)

4. When the immune system is exposed to foreign cells it will mount a defense that includes the production of _____ by _____. (7.3)

5. Antibodies float freely in the _____ and _____. (7.3)

6. When antibodies encounter a foreign antigen they will bind to it, forming a(n) _____. (7.3)

7. Throughout life, individuals carry circulating _____ against any antigens different from their own. (7.3)

8. When red blood cells with foreign antigens are attacked by _____, they clump together or _____. (7.3)

9. Clumping together of red blood cells (RBCs) may lead to _____ or _____. (7.3)

10. An adverse effect of a blood transfusion is called a
    _____. (7.3)

11. To insure that a blood transfusion is safe, blood typing and
    _____ will be done. (7.3)

Short Answer

12. An individual has Type B blood. What type of antigen is found on their red blood cells? What type of antibodies do they carry?

13. An individual has Type O blood. What type of antigen is found on their red blood cells? What type of antibodies do they carry?

14. Two individuals both have blood type A; however, one is type A+ and the other is type A−. What is the difference in their red blood cells?

15. Why can an individual with Rh− blood receive a transfusion of Rh+ blood one time without complications, while a second transfusion of Rh+ blood may cause a severe reaction?

16. (a) What is hemolytic disease of the newborn? (b) What parental blood types would put a newborn at risk for this disease? (c) Why does this disease seldom affect a first child? (d) How can it be prevented in subsequent children?

17. Describe the process of cross-matching. What results indicate that two blood samples are a good match?

7.4 Blood disorders

Matching

___ 1. **carbon monoxide poisoning** a. infection of lymphocytes caused by the Epstein-Barr virus

___ 2. **mononucleosis** b. uncontrolled production of abnormal or immature white blood cells

___ 3. **anemia** c. a reduction of the number of platelets in the blood

___ 4. **multiple myeloma** d. a reduction of the oxygen-carrying capacity of the blood due to blockage of hemoglobin's oxygen-binding sites

___ 5. **leukemia** e. a reduced ability of the blood to transport oxygen due to a variety of causes

___ 6. **blood poisoning** f. uncontrolled division of abnormal plasma cells in the bone marrow

___ 7. **thrombocytopenia** g. proliferation of microorganisms in the blood plasma

Chapter Test

Multiple Choice

1. The circulatory system consists of:
    a. red blood cells, white blood cells, and platelets.
    b. plasma and formed elements.
    c. blood vessels and lymphatic vessels.
    d. heart, blood vessels, and blood.

2. Which of the following is not a function of the blood?
    a. protection against infection and illness
    b. regulation of body temperature and fluid balance
    c. secretion of erythropoietin
    d. transportation

3. Formed elements include all of the following *except:*
    a. red blood cells.
    b. white blood cells.
    c. plasma.
    d. platelets.

4. Albumins are a type of:
    a. blood cell.
    b. platelet.
    c. plasma protein.
    d. hormone.

5. LDLs are:
   a. a beneficial type of cholesterol.
   b. a type of clotting protein.
   c. a lipoprotein associated with heart disease.
   d. cell surface antigens.

6. The hemoglobin molecule of red blood cells consists of:
   a. four polypeptide chains, each containing a heme group with an iron atom.
   b. four heme chains, each containing an iron group with a central protein.
   c. four iron atoms surrounding a central, iron-binding protein.
   d. two polypeptide chains, two heme groups, and two iron atoms.

7. Deoxyhemoglobin:
   a. contains less iron than oxyhemoglobin.
   b. is bright red.
   c. has given up its oxygen.
   d. carries less carbon dioxide than oxyhemoglobin.

8. Red blood cells are accurately described as:
   a. large and oblong cells.
   b. rigid and inflexible cells.
   c. small and flat cells.
   d. multinucleate cells.

9. Stem cells in the bone marrow produce:
   a. red blood cells only.
   b. white blood cells only.
   c. red and white blood cells.
   d. red cells, white cells, and platelets.

10. Red blood cells have a short life span because:
    a. they are easily damaged as they pass through the blood vessels.
    b. they are often destroyed by immune system cells.
    c. they lack a nucleus.
    d. their cell membranes are impermeable to required nutrients.

11. Bilirubin is produced during:
    a. the conversion of prothrombin into thrombin.
    b. the synthesis of erythropoietin.
    c. the differentiation of macrophages.
    d. the breakdown of heme.

12. Erythropoietin stimulates production of _____ in response to _____.
    a. white blood cells, infection
    b. red blood cells, low O_2 availability
    c. platelets, blood vessel damage
    d. fibrin, increased fibrinogen levels

13. Neutrophils:
    a. defend the body against large parasites.
    b. release histamine when tissues are injured.
    c. are classified as B cells or T cells.
    d. are the first white blood cells to combat infection.

14. All of the following are granular leukocytes *except:*
    a. neutrophils.
    b. monocytes.
    c. eosinophils.
    d. basophils.

15. Which of the following conditions would increase the release of O_2 from red blood cells into the tissue fluids?
    a. neutral pH
    b. high concentration of O_2 in the tissue fluids
    c. low concentration of O_2 in the tissue fluids
    d. low concentration of CO_2 in the tissue fluids

16. In the process of hemostasis, when a blood vessel breaks:
    a. red blood cells differentiate into macrophages.
    b. eosinophils migrate to the site to seal the rupture.
    c. platelets swell, develop spiky extensions, and begin to clump together.
    d. the kidneys decrease production of erythropoietin.

17. In what order do the following substances participate in the formation of a blood clot?
    I. fibrinogen
    II. thrombin
    III. prothrombin activator
    IV. fibrin
    V. prothrombin
    a. IV, I, V, III, II
    b. I, IV, III, V, II
    c. II, V, III, IV, I
    d. III, V, II, I, IV

18. In the past, some hemophiliacs contracted AIDS because:
    a. blood transfusions transmitted HIV antibodies.
    b. frequent hospital stays increased exposure to HIV.
    c. reduced immune function increased susceptibility to AIDS.
    d. clotting factor VIII could not be screened for HIV.

19. Agglutination between red blood cells of two blood samples indicates:
    a. the two samples are compatible.
    b. the cells of both sample have the same antigens.
    c. the two samples are not compatible.
    d. the antigens of one sample have attacked the antibodies of the other sample.

20. During a transfusion reaction:
    a. recipient antigens attack donor antibodies.
    b. recipient antigens fail to bind to donor antigens.
    c. recipient antibodies attack donor antigens.
    d. recipient antibodies engulf and destroy foreign red blood cells by phagocytosis.

21. In the ABO blood system, antibodies appear in an individual's blood:
    a. only in response to foreign antigens.
    b. early in life, with or without exposure to foreign antigens.
    c. only during childhood.
    d. only in response to Rh+ blood.

22. Early symptoms of carbon monoxide poisoning include:
    a. flushed skin.
    b. headache.
    c. nausea.
    d. all of the above.

23. Which type of anemia results from a deficiency in vitamin B_{12} absorption by the digestive tract?
    a. iron deficiency anemia
    b. sickle cell anemia
    c. aplastic anemia
    d. pernicious anemia

24. Mononucleosis:
    a. is not contagious.
    b. is the result of a bacterial infection.
    c. has no known cure.
    d. symptoms include easy bruising and bone pain.

25. An individual with blood type A+ could receive a transfusion of blood type _____.
    a. B−
    b. AB+
    c. A−
    d. AB−

Key Concept Review Questions

Each of the Key Concepts listed at the beginning of this chapter has been rewritten as a question below. After successively completing the study guide exercises and the Chapter Test, you should be able to answer each of these questions. Refer to the Key Concepts list on the first page of this chapter to check your answers.

1. What are the functions of the blood?
2. What are the two main components of blood?
3. What is plasma composed of?
4. What are three important categories of plasma proteins and their functions?
5. What are the formed elements of the blood?
6. Where are formed elements produced? What type of cell is involved?
7. What is found inside red blood cells? What is the function of hemoglobin?
8. Where and how are old red blood cells destroyed?
9. What hormone controls the production of red blood cells?
10. What is the function of white blood cells? How are they classified?
11. Which white blood cells are classified as granular leukocytes? Which white blood cells are classified as agranular leukocytes?
12. What process requires platelets? What hormone regulates platelet production?
13. What is hemostasis? What are the stages of hemostasis?
14. What substances are required for the formation of a blood clot?
15. What determines blood compatibility?
16. What determines an individual's blood type?
17. What are two commonly used blood typing systems?
18. What causes a transfusion reaction?
19. What parts of the blood may be involved in blood disorders?
20. What are seven blood disorders discussed in Chapter 7?

Heart and Blood Vessels

Chapter Summary and Key Concepts

After reading and studying this chapter you should know the following:

Sections 8.1, 8.2

1. The pumping action of the heart provides the power to transport blood.
2. Arteries transport blood away from the heart; veins transport blood toward the heart.
3. Capillaries transport blood from arteries to veins, and they allow for exchange of fluid, nutrients, and wastes with the tissues.
4. The structure of the vessel wall is suited to its function.
5. Veins return blood to the heart with the aid of skeletal muscle activity, one-way valves, and pressure changes associated with breathing.
6. The lymphatic system helps to maintain blood volume by returning excess extracellular fluid to the blood.
7. The heart consists of four chambers—two upper atria that receive blood, and two lower ventricles that eject blood from the heart.
8. Heart valves prevent blood from flowing backward.
9. The right side of the heart receives and then pumps deoxygenated blood. The left side of the heart receives and then pumps oxygenated blood.
10. The pulmonary circuit transports blood between the heart and the lungs for oxygenation.
11. The systemic circuit transports blood between the heart and the rest of the body to supply oxygen and nutrients to the tissues and remove wastes from the tissues.
12. The cardiac cycle includes periods of contraction called systole, and periods of relaxation called diastole.

13. Electrical impulses cause contraction of heart muscle, and they are initiated and spread through the heart by the cardiac conduction system.

Sections 8.3, 8.4, 8.5, 8.6

14. Blood pressure is the force that blood exerts on the vessel walls.
15. Blood pressure is an important indicator of the overall health of the cardiovascular system.
16. Systolic pressure is produced when ventricles contract, and diastolic pressure is produced when ventricles relax.
17. A constant arterial blood pressure is crucial to homeostasis, and is maintained by regulating heart rate, force of contraction of the heart, and the diameter of arterioles.
18. Regulation of the cardiovascular system is accomplished by receptors in the arteries, regulatory centers in the brain, hormones, and the environment in the tissues.
19. Cardiovascular disorders include hypertension, angina pectoris, heart attack, congestive heart failure, embolism, and stroke.
20. There are many voluntary actions that can help to maintain cardiovascular health.

Exercises

Complete the exercises for each section after you have read and studied the section. If you cannot answer some questions, or answer them incorrectly, return to the chapter and review this information. You may find it helpful to work on only one section at a time. When you have completed all sections, take the Chapter Test as an indicator of your mastery of this topic.

8.1 Blood vessels transport blood

8.2 The heart pumps blood through the vessels

Matching

____ 1. **cardiovascular system**   a. band of smooth muscle that controls blood flow into the capillaries

____ 2. **arteries**   b. the outermost layer of the heart

____ 3. **arterioles**   c. valves that prevent blood from flowing back into the atria when the ventricles contract

____ 4. **precapillary sphincter**   d. pathway of blood flow that sends deoxygenated blood to the lungs and returns oxygenated blood to the heart

____ 5. **capillaries**   e. the inner layer of the heart

_____ 6. **veins** f. period of relaxation of cardiac muscle

_____ 7. **epicardium** g. the layer of the heart wall consisting mainly of cardiac muscle

_____ 8. **myocardium** h. period of contraction in the cardiac muscle

_____ 9. **endocardium** i. record of electrical impulses in the cardiac conduction system

_____ 10. **atria** j. vessels that transport blood from the body back to the heart

_____ 11. **ventricles** k. sinoatrial (SA) node

_____ 12. **septum** l. specialized cardiac muscle cells that initiate and distribute electrical impulses through the heart

_____ 13. **atrioventricular valves (AV)** m. the largest artery in the body

_____ 14. **semilunar valves** n. pathway of blood flow that carries oxygenated blood to the body and returns deoxygenated blood to the heart

_____ 15. **pulmonary circuit** o. the two upper chambers of the heart

_____ 16. **systemic circuit** p. body system consisting of the heart and blood vessels

_____ 17. **pulmonary veins** q. blood vessels that supply the heart muscle

_____ 18. **aorta** r. the smallest arteries of the body

_____ 19. **systole** s. the muscular partition separating the right and left sides of the heart

_____ 20. **diastole** t. the smallest vessels whose walls consists of a single layer of cells

_____ 21. **cardiac conduction system** u. vessels that carry oxygenated blood to the heart from the lungs

_____ 22. **cardiac pacemaker** v. thick-walled vessels that transport blood at high pressure away from the heart

_____ 23. **electrocardiogram (ECG)** w. valves that prevent blood from flowing backward into the ventricles when the heart relaxes

_____ 24. **coronary arteries** x. the two lower chambers of the heart

Fill-in-the-Blank

Referenced sections are indicated in parentheses.

25. The three major types of blood vessels in the body are _____, _____, and _____. (8.1)

26. The elastic tissue in arteries allows them to _____ under pressure. (8.1)

27. Large arteries are able to withstand high pressure because the artery wall contains a thick layer of _____. (8.1)

continued

28. The hollow interior of a blood vessel is the _____. (8.1)

29. The _____ of a blood vessel is a continuation of the lining of the heart. (8.1)

30. The middle layer of the arterial vessel wall is composed primarily of _____ _____. (8.1)

31. Water and small solutes escape from a capillary through _____ in the epithelial cells and _____ between the cells. (8.1)

32. Ballooning of the arterial wall due to vessel damage is referred to as a(n) _____. (8.1)

33. Contraction of vascular smooth muscle is called _____. (8.1)

34. Arterioles connect directly to _____. (8.1)

35. Red blood cells often must pass through capillaries _____ _____. (8.1)

36. _____ _____ are extensive networks of capillaries found in all areas of the body. (8.1)

37. Filtration of fluid out of the capillaries is caused by _____ _____ generated by the heart. (8.1)

38. Blind-ended vessels that pick up excess extracellular fluid are _____ _____. (8.1)

39. The fluid carried by lymphatic vessels is called _____. (8.1)

40. Blood flows from capillaries into _____. (8.1)

41. Veins permanently swollen by pooled blood are called _____ veins. (8.1)

42. The _____ is a tough fibrous sac surrounding the heart. (8.2)

43. An abnormality in the heart rate is called a(n) _____. (8.2)

44. The buildup of fatty deposits in the arteries is called _____. (8.2)

Short Answer

45. List three functions of arterioles.

46. How does the structure of a capillary wall enable blood to exchange substances with tissue cells?

47. Why do the walls of veins not require as much muscle strength as the walls of arteries?

48. What three mechanisms help the veins return blood to the heart?

49. List the four structures of the cardiac conduction system.

Paragraph Completion

Use the following terms to complete the paragraph, and then repeat this exercise with the terms covered. Terms used more than once are listed multiple times.

SA node	aorta	atrium	atrial systole
diastole	diastole	systole	pulse
AV node	atria	AV	systole
diastole	conduction	ventricular systole	atria
heart sounds	AV bundle	pulmonary trunk	
ventricles	brain	ventricles	

A complete cardiac cycle involves contraction of the two (50)_____, which forces blood into the ventricles, followed by contraction of the two (51)_____, which pumps blood into the pulmonary arteries and the aorta. The period of contraction is referred to as (52)_____, and the period of relaxation is referred to as (53)_____. The cardiac cycle begins with (54)_____, when all chambers of the heart are relaxed. Blood flows from the veins into the (55)_____ and then passively into the

continued

(56)_____. The next stage is (57)_____ when the atria contract, forcing any remaining blood into the ventricles. The final stage is (58)_____, which closes the two (59)_____ valves and forces blood into the (60)_____ and (61)_____. Artery walls stretch during (62)_____ and recoil during (63)_____; this can be felt in arteries near the skin surface as a (64)_____. The closing of the valves during the cardiac cycle generates (65)_____. Coordination of the cardiac cycle is due to the cardiac (66)_____ system. Electrical impulses are initiated in the (67)_____ located in the right (68)_____. Impulses travel across the heart, next reaching the (69)_____. In the septum between the two ventricles lie conducting fibers called the (70)_____, which branch and carry the electrical impulses throughout the ventricles. In addition to impulses initiated by the heart's own pacemaker, heart rate can be influenced by messages from the (71)_____.

Ordering

Order the following structures to indicate the correct sequence of blood flow, beginning and ending with the right atrium.

left atrioventricular valve	venules	veins
right ventricle	right atrioventricular valve	pulmonary trunk
aortic semilunar valve	pulmonary semilunar valve	arterioles
lungs	capillaries	pulmonary veins
aorta	left atrium	arteries
		left ventricle

72. right atrium →(a)_____ →(b)_____

→(c)_____ →(d)_____ →(e)_____

→(f)_____ →(g)_____ →(h)_____

→(i)_____ →(j)_____ →(k)_____

→(l)_____ →(m)_____ →(n)_____

→(o)_____ →(p)_____ →right atrium

Labeling

73. Label Figure 8.1 with the words listed below.

pulmonary trunk   pulmonary semilunar valve
aortic semilunar valve   right ventricle
left AV valve   right AV valve
aorta   septum
right atrium   left ventricle
left atrium

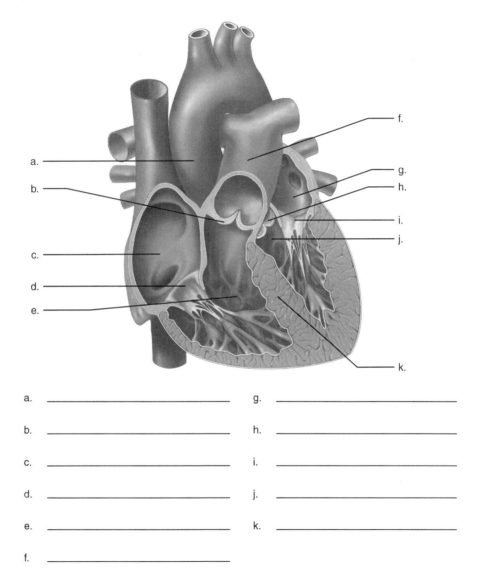

a. _____   g. _____

b. _____   h. _____

c. _____   i. _____

d. _____   j. _____

e. _____   k. _____

f. _____

Figure 8.1

Labeling

In Figure 8.2, indicate one answer from each column that corresponds to the electrical pattern of the ECG.

ECG	Electrical Activity	Muscle Activity
a. QRS complex	d. electrical impulse travels across the atria	g. immediately precedes ventricular contraction
b. T wave	e. end of electrical activity in the ventricles	h. atrial contraction
c. P wave	f. spread of electrical impulse down the septum and around the ventricles	i. contraction cycle has ended

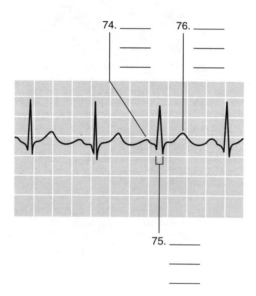

Figure 8.2

8.3 **Blood exerts pressure against vessel walls**

8.4 **How the cardiovascular system is regulated**

8.5 **Cardiovascular disorders: A major health issue**

8.6 **Reducing your risk of cardiovascular disease**

Matching

____ 1. **blood pressure** a. special structures in the arteries that monitor and help to regulate blood pressure

____ 2. **systolic pressure** b. death of heart tissue due to lack of oxygen

____ 3. **diastolic pressure** c. the lowest blood pressure, occurs during ventricular relaxation

____ 4. **hypertension** d. the amount of blood the heart pumps into the aorta in 1 minute

____ 5. **baroreceptors** e. damage to brain tissue caused by interruption of the blood supply

____ 6. **cardiac output** f. buildup of extracellular fluid due to weakening of heart muscle

____ 7. **angina pectoris** g. sudden blockage of a blood vessel, often due to a blood clot

____ 8. **heart attack** h. high blood pressure

____ 9. **congestive heart failure** i. force exerted by blood on the walls of the blood vessels

____ 10. **embolism** j. sensations of pain and tightness in the chest

____ 11. **stroke** k. highest blood pressure, occurs during ventricular contraction

Fill-in-the-Blank

Referenced sections are indicated in parentheses.

12. Blood pressure is measured with a _____. (8.3)

13. In measuring blood pressure, the pressure at which blood reenters the collapsed artery is the _____ pressure. (8.3)

14. When diastolic pressure remains normal while systolic pressure is too high, this is called _____ _____. (8.3)

15. _____ is a condition when blood pressure falls below normal. (8.3)

16. Cardiac output is determined by multiplying _____ _____ by _____ _____. (8.4)

continued

17. _____ nerves from the _____ _____ of the brain can stimulate the heart to beat faster. (8.4)

18. Cardiovascular disorders are the number _____ killer in the United States. (8.5)

19. A(n) _____ is an x-ray of the blood vessels. (8.5)

20. _____ is a surgical procedure in which blood vessels are taken from one part of the body and grafted onto a blocked coronary artery. (8.5)

Short Answer

Answer the following questions about Figure 8.3.

21. (a) What is represented by the highest peaks (120 mm Hg) in the arteries?

    (b) What is happening in the heart during this time?

22. (a) What is represented by the lowest peaks (80 mm Hg) in the arteries?

    (b) What is happening in the heart during this time?

23. Why does blood pressure decrease in the arterioles and capillaries?

24. How does blood flow in the capillaries differ from blood flow in the arteries?

25. How do blood pressure differences in the arteries, capillaries, and veins keep blood flowing through the body?

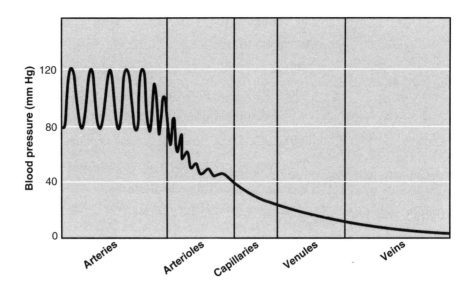

Figure 8.3

Paragraph Completion

Use the following terms to complete the paragraph, and then repeat this exercise with the terms covered.

arterial	brain	parasympathetic	carotid arteries
increase	heart rate	epinephrine	sympathetic
dilate	oxygen	carbon dioxide	negative feedback
diameter	aorta	adrenal glands	medulla oblongata

Homeostatic regulation of the cardiovascular system depends on maintaining a constant (26)_____ blood pressure. This regulation is accomplished by the activity of several structures. Baroreceptors, located in the (27)_____ and (28)_____, stretch as blood pressure rises. This sends signals to the cardiovascular center in the (29)_____. The response is lowering of the (30)_____, which reduces cardiac output. Signals received by the brain in response to falling blood pressure trigger a(n) (31)_____ in cardiac output and constriction of the arterioles. This activity between the baroreceptors and the brain represents a (32)_____ loop. The cardiovascular center is located in the (33)_____ of the brain. Nerves from this area can influence heart activity. (34)_____ nerves stimulate the heart and increase heart rate, while (35)_____ nerves inhibit the heart. These nerves can also affect blood pressure

continued

by changing the (36)_____ of the arterioles. The hormone (37)_____ produced by the (38)_____ can stimulate the heart in times of stress. Blood flow in a particular area is regulated by local tissue needs. When tissues are metabolically active, they consume more (39)_____ and produce more (40)_____. This causes precapillary sphincters to (41)_____, increasing blood flow to the area.

Chapter Test

Word Choice

Circle the accurate term to complete each sentence.

1. Vasodilation of arterioles (increases/decreases) their diameter, (increasing/decreasing) blood flow to the capillaries.

2. The walls of veins are (thinner/thicker) than the walls of arteries.

3. The lumen of a vein is (larger/smaller) in diameter than the lumen of an artery.

4. Blood pressure in the veins is (less than/greater than) blood pressure in the arteries.

5. During diastole, pressure inside the heart chambers is (less than/greater than) blood pressure in the veins.

6. Baroreceptors stretch as blood pressure (increases/decreases).

7. Blood entering the right atrium is (oxygenated/deoxygenated).

Multiple Choice

8. Which chamber of the heart is the most muscular?
    a. right atrium
    b. left atrium
    c. right ventricle
    d. left ventricle

9. In the functioning of the cardiovascular system, the heart provides _____, while the blood vessels provide _____.
   a. new red blood cells, oxygen
   b. baroreceptors to monitor blood pressure, smooth muscle capable of changing vessel diameter to respond to blood pressure
   c. power to move the blood, a network through which blood can flow
   d. hormones to regulate cardiac output, receptors to monitor cardiac output

10. Large- and medium-sized arteries are:
    a. stiff, yet elastic.
    b. thin, yet strong.
    c. thick, yet diffusible.
    d. none of the above.

11. Which layer of a large blood vessel's wall will be primarily smooth muscle?
    a. inner layer
    b. middle layer
    c. outer layer
    d. epithelial layer

12. Which of the following accurately describes arterioles?
    a. they have the same outer layer of connective tissue as that found in arteries
    b. they deliver blood directly to a venule
    c. they help regulate the amount of blood that flows to each capillary
    d. blood pressure in the cardiovascular system is highest in the arterioles

13. Capillary walls consist of:
    a. a thin layer of smooth muscle surrounding the endothelium.
    b. smooth muscle enclosed by a thin layer of connective tissue.
    c. a single layer of endothelium composed of epithelial cells.
    d. three layers similar to larger arteries, but each layer is much thinner.

14. Excess extracellular fluid accumulates in the tissues because:
    a. the capillary wall is too thick to allow the fluid to diffuse back into the vessels.
    b. the extracellular fluid contains too many proteins to allow return to the capillary.
    c. the pressure-induced filtration of water is greater than the diffusional reabsorption of water.
    d. the extracellular fluid provides storage for excess oxygen.

15. Almost two-thirds of all the blood in the body is:
    a. in the heart.
    b. in the veins.
    c. in the arteries.
    d. in the capillary beds.

16. Which of the following would interfere most with the ability of the veins to return blood to the heart?
    a. absence of valves in the veins
    b. low blood pressure in the veins
    c. high blood pressure in the arteries
    d. dilation of precapillary sphincters

17. The protective sac that encases the heart and anchors it to surrounding structures is the:
    a. pericardium.
    b. endocardium.
    c. endothelium.
    d. myocardium.

18. Atrioventricular valves prevent blood from flowing backward into the:
    a. atria.
    b. ventricles.
    c. pulmonary veins.
    d. aorta.

19. The movement of blood through the pulmonary circuit transports:
    a. oxygenated blood from the lungs directly to the body.
    b. oxygenated blood from the lungs to the heart.
    c. deoxygenated blood from the lungs to the heart.
    d. deoxygenated blood from the left side of the heart to the right side of the heart.

20. The ventricles fill with blood passively during:
    a. atrial systole.
    b. ventricular systole.
    c. atrial and ventricular diastole.
    d. semilunar diastole.

21. Which of the following represents the correct order in which electrical impulses will pass through the heart?
    a. AV node, AV bundle, SA node, Purkinje fibers
    b. SA node, Purkinje fibers, AV node, AV bundle
    c. AV bundle, AV node, SA node, Purkinje fibers
    d. SA node, AV node, AV bundle, Purkinje fibers

22. Which of the following would be occurring in a patient who lacked a P wave?
    a. no contraction of the atria
    b. no relaxation of the atria
    c. no contraction of the ventricles
    d. no relaxation of the ventricles

23. Shortness of breath, swollen ankles, and weight gain are symptoms of:
    a. hypertension.
    b. hypotension.
    c. cerebral embolism.
    d. congestive heart failure.

24. An increase in cardiac output would result from all of the following *except:*
    a. epinephrine.
    b. stimulation by sympathetic nerves.
    c. a decrease in stroke volume.
    d. an increase in heart rate.

25. If the cardiovascular center in the brain senses a drop in blood pressure, which of the following would occur?
    a. Heart rate would be lowered.
    b. Cardiac output would be reduced.
    c. Vasoconstriction of the arterioles would occur.
    d. Parasympathetic nerves would carry signals to the heart.

Key Concept Review Questions

Each of the Key Concepts listed at the beginning of this chapter has been rewritten as a question below. After successfully completing the study guide exercises and the Chapter Test, you should be able to answer each of these questions. Refer to the Key Concepts list to check your answers.

1. What provides the power to transport blood?
2. Which vessels transport blood away from the heart? Which vessels transport blood toward the heart?
3. What is the function of the capillaries?
4. What is important about the structure of blood vessel walls?
5. What three mechanisms help veins return blood to the heart?
6. How does the lymphatic system help to maintain blood volume?
7. How would you describe the four chambers of the heart?

8. What is the function of the heart valves?
9. What type of blood enters and then leaves each side of the heart?
10. What is the function of the pulmonary circuit?
11. What is the function of the systemic circuit?
12. How would you describe the periods of the cardiac cycle?
13. What is the role of the cardiac conduction system?
14. What is blood pressure?
15. Why is blood pressure important?
16. What produces systolic pressure? What produces diastolic pressure?
17. What three variables must be regulated to maintain a constant arterial blood pressure?
18. What factors contribute to regulation of the cardiovascular system?
19. What are six disorders of the cardiovascular system discussed in Chapter 8?
20. How much effect can personal lifestyle choices have on cardiovascular health?

The Immune System and Mechanisms of Defense

Chapter Summary and Key Concepts

After reading and studying this chapter you should know the following:

Sections 9.1, 9.2, 9.3, 9.4, 9.5

1. The immune system is a group of cells, proteins, and structures of the lymphatic and circulatory systems that defend the body.

2. Pathogens are microorganisms that cause cell death and disease.

3. The danger of a particular pathogen is determined by its transmissibility, mode of transmission, and virulence.

4. The lymphatic system consists of a series of vessels, nodes, and organs that transport and cleanse the lymph.

5. The three lines of defense of the body are physical and chemical barriers, nonspecific defense mechanisms, and specific defense mechanisms.

6. The first line of defense, consisting of physical and chemical barriers, includes skin, tears, saliva, earwax, digestive acids, mucus, vomiting, urination, defecation, and resident bacteria.

7. The second line of defense consists of nonspecific mechanisms that attack foreign and damaged cells, including phagocytosis by neutrophils and macrophages, destruction of large invaders by eosinophils, the inflammatory response, natural killer cells, the complement system, interferons, and fever.

8. The third line of defense involves specific mechanisms that target specific pathogens for destruction, and depends on B cells and T cells.

Sections 9.6, 9.7, 9.8, 9.9

9. The three important characteristics of the immune response are (1) it recognizes and targets specific pathogens, (2) it has memory, and (3) it protects the entire body.

10. An immune response is any response utilizing the components of the immune system. An antigen is any substance that can provoke an immune response.

11. Body cells are normally protected from the immune response by the presence of major histocompatibility complex (MHC) proteins.

12. Lymphocytes called B cells are responsible for antibody-mediated immunity. B cells produce antibodies capable of destroying a particular antigen.

13. Lymphocytes called T cells are responsible for cell-mediated immunity. T cells exist in several types and function to stimulate other cells and destroy infected body cells.

14. B cells recognize and bind to free antigens. T cells can only recognize and bind to an antigen that is presented by an antigen-presenting cell.

15. Both B cells and T cells produce memory cells that trigger the immune response faster when activated a second time. Memory cells are responsible for immunity.

16. Active immunization occurs when the body is exposed to a vaccine.

17. Passive immunization occurs when a person receives prepared antibodies for a specific pathogen.

18. Antibiotics are only effective against bacteria.

Sections 9.10, 9.11

19. Disorders of the immune system include allergies, lupus erythematosus, rheumatoid arthritis, and AIDS.

20. AIDS results from infection by the human immunodeficiency virus (HIV), and it causes a reduction in the number of helper T cells.

Exercises

Complete the exercises for each section after you have read and studied the section. If you cannot answer some questions, or answer them incorrectly, return to the chapter and review this information. You may find it helpful to work on only one section at a time. When you have completed all sections, take the Chapter Test as an indicator of your mastery of this topic.

9.1   **Pathogens cause disease**

9.2   **The lymphatic system defends the body**

9.3   **The body has three lines of defense**

9.4   **Physical and chemical barriers: Our first line of defense**

9.5   **Nonspecific defenses: The second line of defense**

Matching

____ 1. **immune system**   a. the largest lymphatic organ, responsible for removing old red blood cells from circulation and fighting infections

____ 2. **pathogens**   b. a group of white blood cells that is nonspecific in destroying tumor cells and infected body cells

____ 3. **bacteria**   c. abnormally high body temperature

____ 4. **virus**   d. white blood cells that cluster around large invaders and destroy them by releasing digestive enzymes

____ 5. **lymphatic system**   e. a process in which white blood cells engulf and destroy foreign cells

____ 6. **lymph nodes**   f. connective tissue cells stimulated by the inflammatory response to release histamine

____ 7. **spleen**   g. a group of cells, proteins, and structures of the lymphatic and circulatory systems that defend the body

____ 8. **thymus gland**   h. prokaryotic organisms

____ 9. **lysozyme**   i. the first white blood cells to respond to an infection

____ 10. **phagocytosis**   j. a group of 20 plasma proteins that assist defense mechanisms

____ 11. **neutrophils**   k. monocytes that develop into large phagocytic cells

____ 12. **macrophages**   l. body system responsible for maintaining blood volume, transporting fats, and defending the body

____ 13. **eosinophils**   m. site of maturation of T cells

____ 14. **inflammation**   n. white blood cells that secrete histamine

____ 15. **mast cells**   o. an enzyme found in tears and saliva that kills many bacteria

____ 16. **histamine**   p. small organs of the lymphatic system that remove microorganisms, debris, and abnormal cells from the lymph

____ 17. **basophils**   q. the smallest known infectious agent

____ 18. **natural killer cells**   r. microorganisms that cause cell death and disease

____ 19. **complement system**   s. a protein secreted by virus-infected cells that diffuses to healthy neighboring cells and aids protection

____ 20. **interferon**   t. a chemical that promotes vasodilation of small blood vessels

____ 21. **fever**   u. a response to tissue injury that involves redness, warmth, swelling, and pain

Paragraph Completion

Use the following terms to complete the paragraph, and then repeat this exercise with the terms covered.

cardiovascular	lymphatic	macrophages
lymphocytes	thoracic	nodes
subclavian	cells	cardiovascular
fluid	vessels	lymph
neck	capillaries	

The lymphatic system consists of lymphatic vessels, lymph nodes, and associated organs and structures. Lymphatic vessels begin as blind-ended (22) _____, and are closely associated with the (23) _____ system. Lymph capillaries have overlapping (24) _____ and can take up (25) _____, small substances, and bacteria from the tissue spaces. Lymph capillaries merge to form lymphatic (26) _____, which flow into lymph (27) _____. Lymph nodes are clustered in the area of the digestive tract, armpit, groin, and (28) _____, and filter the lymph before returning it to the (29) _____ system. Inside each node are (30) _____ that destroy foreign cells by phagocytosis, and (31) _____ that activate specific defense mechanisms. Filtered (32) _____ fluid will eventually be returned to the blood through the right (33) _____ duct or the (34) _____ duct into the (35) _____ veins.

Labeling

Label the indicated structures in Figure 9.1 on page 117. Indicate both (a) the structure, and (b) the function.

Chapter 9 *The Immune System and Mechanisms of Defense* **117**

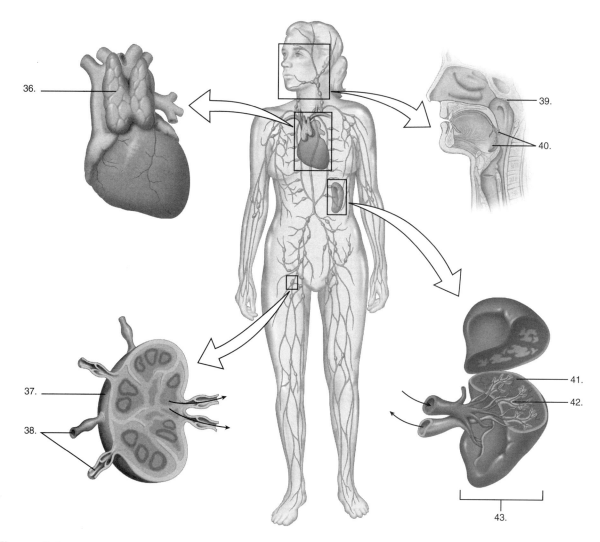

Figure 9.1

36. a. _____   40. a. _____
    b. _____   b. _____
37. a. _____   41. a. _____
    b. _____   b. _____
38. a. _____   42. a. _____
    b. _____   b. _____
39. a. _____   43. a. _____
    b. _____   b. _____

9.6 Specific defense mechanisms: The third line of defense

9.7 Immune memory creates immunity

9.8 Medical assistance in the war against pathogens

9.9 Tissue rejection: A medical challenge

Crossword Puzzle: Defense of the Body

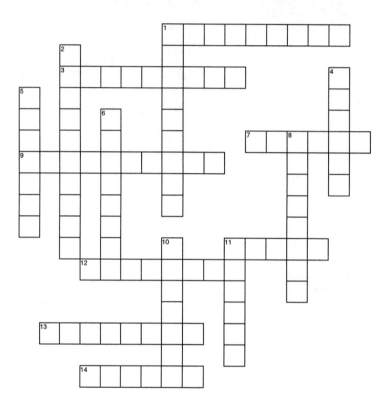

Across

1. T cells with CD8 receptors are activated to produce _____ T cells
3. Tissue _____ is minimized with the use of immunosuppresive drugs
7. Some activated B cells differentiate into _____ cells that produce antibodies
9. _____ antibodies are pure preparations specific for a single antigen
11. Self-markers on body cells are known as _____ histocompatability complex proteins
12. The region of an antibody where the amino acid sequence varies
13. A protein released by cytotoxic T cells that makes holes in the target cell membrane
14. Both B cells and cytotoxic T cells require stimulation by _____ T cells

Down

1. A class of signaling molecules secreted by helper T cells
2. Antigen-_____ cells display fragments of antigen on their surface
4. T cells are lymphocytes that mature in this gland
5. A first exposure to a particular antigen generates a _____ immune response
6. Antibiotics are effective against these microorganisms
8. B cells are responsible for _____-mediated immunity
10. Active immunization occurs when this is administered
11. The presence of these cells is the basis of immunity

Labeling

Use the terms and processes listed below to label Figure 9.2.

a. B cell is exposed to antigen
b. formation of plasma cells
c. B-cell surface receptors
d. formation of memory cells
e. activation
f. mature inactive B cell
g. surface antigen
h. clone formation
i. pathogen
j. plasma cells secrete antibodies into circulation
k. antibodies

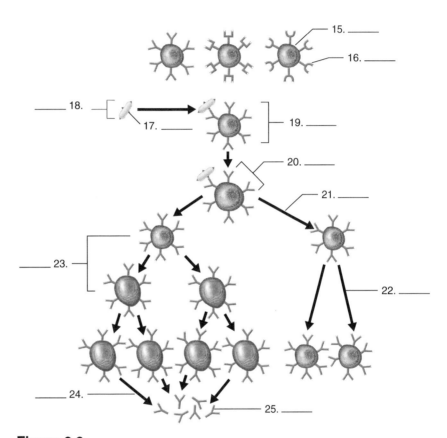

Figure 9.2

Completion

Write the name of each type of antibody described in the table below. Indicate what percentage of immunoglobulins each antibody type comprises.

Type of Antibody	% of Immunoglobulins	Function of Antibody
26.	27.	enters areas of the body covered by mucous membranes
28.	29.	activates the inflammatory response, involved in allergies
30.	31.	long-lived antibodies, can cross the placenta
32.	33.	function unclear, may activate B cells
34.	35.	first antibodies released during an immune response

9.10 Inappropriate immune system activity causes problems

9.11 Immune deficiency: The special case of AIDS

Fill-in-the-Blank

Referenced sections are in parentheses.

1. _____ is an inappropriate response of the immune system to a nondangerous pathogen. (9.10)

2. Anaphylactic shock is a severe _____ allergic reaction. (9.10)

3. _____ disorders result when the immune system targets "self" cells. (9.10)

4. Phase _____ of AIDS is characterized by a gradual decline in the number of T cells and an increased vulnerability to opportunistic infections. (9.11)

Labeling

Use the steps of HIV infection listed below to label the steps in Figure 9.3.

a. the host cell makes a second strand of DNA to complement the first single strand of DNA

b. HIV enters the host cell

c. the host cell expresses the HIV genes and makes new viral RNA and proteins

d. using HIV enzymes, the host cell makes a single-stranded DNA copy of the viral RNA strand

e. the double-stranded DNA fragment is inserted into the host DNA

f. new HIV particles leave the host cell

g. HIV binds to a host cell CD4 receptor

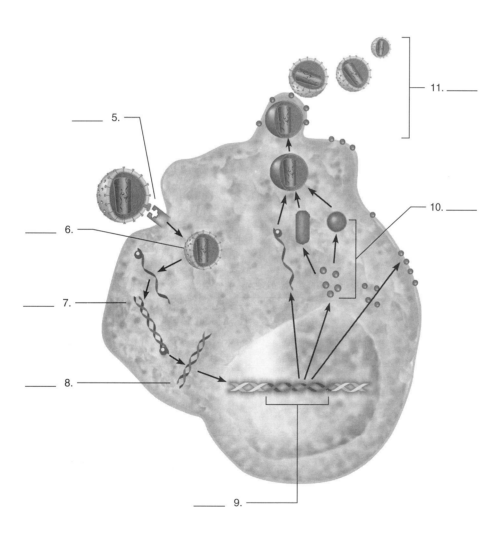

Figure 9.3

Chapter Test

Multiple Choice

1. Microorganisms that cause cell death and disease are called:
   a. antigens.
   b. antibodies.
   c. macrophages.
   d. pathogens.

2. Bacteria:
   a. are always pathogenic.
   b. may be engulfed and destroyed by a macrophage.
   c. are not destroyed by antibodies.
   d. lack antigens on their cell surface.

3. Lymphatic capillaries are well suited to absorb fluid and bacteria from the tissue spaces because their structure includes:
   a. overlapping cells separated by spaces.
   b. smooth muscle that contracts to push fluid into the vessels.
   c. nodes that actively absorb fluid.
   d. highly permeable cells that allow fluid to pass freely.

4. Lymphatic vessels are similar to veins because they both:
   a. allow fluid to flow in two directions.
   b. use skeletal muscle contractions to keep fluids moving.
   c. have a single, unbranched structure.
   d. have the ability to absorb large molecules.

5. The correct path of lymph fluid collected from the tissues and transported to the heart is most accurately represented as:
   a. lymphatic duct → lymph node → lymphatic capillary → lymphatic vessel
   b. lymphatic vessel → lymphatic capillary → lymphatic duct → lymph node
   c. lymph node → lymphatic capillary → lymphatic vessel → lymphatic duct
   d. lymphatic capillary → lymphatic vessel → lymph node → lymphatic duct

6. Lymph nodes cleanse the _____ and the spleen cleanses the _____.
   a. blood, blood
   b. blood, lymph
   c. lymph, blood
   d. lymph, lymph

7. Specific defense mechanisms differ from nonspecific defense mechanisms because only specific mechanisms:
   a. utilize macrophages.
   b. target particular antigens.
   c. protect a localized region.
   d. involve complement proteins.

8. Lysozyme:
   a. is an antibacterial enzyme found in tears and saliva.
   b. is an antibacterial enzyme found in B cells.
   c. is a chemical signal released by damaged macrophages.
   d. is a protective chemical released by infected body cells.

9. The first white blood cells to respond to an infection are:
   a. eosinophils.
   b. macrophages.
   c. neutrophils.
   d. helper B cells.

10. Histamine:
    a. is released by mast cells.
    b. promotes vasoconstriction of blood vessels.
    c. is destroyed by mast cells.
    d. activates B cells and T cells.
    e. protects neighboring cells from viral infection.

11. A group of at least 20 proteins that circulate in the blood and assist defense mechanisms is called:
    a. natural killer cells.
    b. interferons.
    c. complement.
    d. cytokines.

12. Fever is caused when _____ release _____.
    a. macrophages, pyrogens
    b. infected cells, interferon
    c. B cells, cytokines
    d. eosinophils, histamines

13. A substance capable of provoking an immune response is a(n):
    a. antibody.
    b. antigen.
    c. lymphokine.
    d. complement.

14. Major histocompatibility complex proteins:
    a. are found in the cytoplasm.
    b. attract pathogens.
    c. identify a cell as "self."
    d. are produced by B cells.

15. B cells mature in the _____, and T cells mature in the _____.
    a. bone marrow, bone marrow
    b. bone marrow, thymus gland
    c. thymus gland, bone marrow
    d. thymus gland, thymus gland

16. Antibodies are produced by:
    a. plasma cells.
    b. helper T cells.
    c. natural killer cells.
    d. antigens.

17. T cells are responsible for:
    a. antibody-mediated immunity.
    b. cell-mediated immunity.
    c. nonspecific immunity.
    d. passive immunity.

18. When a T cell with CD8 receptors contacts an APC displaying an antigen-MHC complex:
    a. the APC divides to form helper T cells.
    b. the T cell produces a clone of cytotoxic T cells.
    c. the antigen-MHC complex is destroyed.
    d. the CD8 receptors are converted to CD4 receptors.

19. A specific antibody will bind to:
    a. a general class of antigens.
    b. a specific antigen.
    c. any antigen because the heavy chains can accommodate any shape.
    d. any antigen because the binding sites are variable.

20. Which of the following can function as antigen-presenting cells?
    a. natural killer cells
    b. macrophages
    c. mast cells
    d. antibodies

21. A secondary immune response could not occur without:
    a. memory cells.
    b. complement.
    c. neutrophils.
    d. natural killer cells.

22. Administration of prepared antibodies produces:
    a. active immunity.
    b. passive immunity.
    c. delayed immunity.
    d. permanent immunity.

23. The actions and effectiveness of antibiotics depends on differences between bacterial and human cells, including:
    a. bacteria have a thick cell wall, human cells do not.
    b. the membrane-bound organelles of bacteria contain different proteins than those of human cells.
    c. bacteria do not have ribosomes.
    d. human protein synthesis occurs faster than bacterial protein synthesis.

24. HIV interferes with specific immunity because it infects and eventually destroys:
    a. B cells.
    b. cytotoxic T cells.
    c. helper T cells.
    d. MHC cells.

25. Autoimmune disorders occur when:
    a. the immune system loses a large number of T cells.
    b. the ability of B cells to produce antibodies is reduced.
    c. the immune system fails to recognize "self" cells.
    d. macrophages function only in nonspecific immunity.

Key Concept Review Questions

Each of the Key Concepts listed at the beginning of this chapter has been rewritten as a question below. After successfully completing the study guide exercises and the Chapter Test, you should be able to answer each of these questions. Refer to the Key Concepts list at the beginning of this chapter to check your answers.

1. The immune system comprises which molecules, structures, and systems?

2. What are pathogens?

3. What determines the danger of a particular pathogen?
4. The lymphatic system comprises which structures and organs?
5. What are the three lines of defense of the body?
6. Which structures and processes make up the first line of defense?
7. How would you describe the structures and processes that make up the second line of defense?
8. Is the third line of defense specific or nonspecific? What cells are responsible for the third line of defense?
9. What are the three important characteristics of the immune response?
10. What is an immune response? What is an antigen?
11. How are body cells normally protected from the immune system?
12. What type of immunity involves B cells? What molecules do B cells produce?
13. What type of immunity involves T cells? What do T cells do?
14. What type of antigen is recognized by B cells? What type of antigen is recognized by T cells?
15. What process are memory cells responsible for? What cells produce memory cells?
16. What results in active immunity?
17. What results in passive immunity?
18. What pathogens are antibiotics effective against?
19. What are four disorders of the immune system discussed in the chapter?
20. What causes AIDS? What effect does AIDS have on T cells?

10

The Respiratory System: Exchange of Gases

Chapter Summary and Key Concepts

After reading and studying this chapter you should know the following:

Sections 10.1, 10.2

1. The term respiration includes four processes: ventilation, external respiration, internal respiration, and cellular respiration.

2. Ventilation is the process of moving air into and out of the lungs, external respiration is the exchange of gases between inhaled air and the blood, internal respiration is the exchange of gases between the blood and tissue fluids, and cellular respiration is the process of using oxygen to produce ATP within cells.

3. The upper respiratory tract consists of the nose and pharynx, and it functions to filter, warm, and humidify air.

4. The lower respiratory tract consists of the larynx, trachea, bronchi, and lungs, and it functions to move air and exchange gases.

5. The pharynx is a common passageway for food, liquid, and air. Food and liquid are prevented from moving into the air passageways by the epiglottis.

6. The larynx also functions as the voice box. Vibration of the vocal cords produces sound.

7. The trachea is reinforced with incomplete cartilage rings; it branches into the bronchi, which are reinforced by cartilage tissue; the bronchi further branch into the bronchioles, which have no cartilage.

8. Bronchioles terminate in tiny air-filled sacs called alveoli that function as the site of gas exchange.

9. The lungs are in the thoracic cavity and are enclosed by the pleural membranes.

Sections 10.3, 10.4, 10.5, 10.6

10. Contraction of the intercostal muscles and the diaphragm enlarge the thoracic cavity, resulting in enlargement of the lungs.

11. Air pressure within the lungs decreases when lung volume increases.

12. Air moves into the lungs down a pressure gradient.

13. Inspiration is the movement of air into the lungs. It is an active process because it requires muscular effort.

14. Passive relaxation of the intercostals and the diaphragm reduces lung volume, increasing air pressure within the lungs. As the air pressure within the lungs becomes greater than the atmospheric pressure, air rushes out of the lungs.

15. Expiration is the movement of air out of the lungs. It is usually a passive process.

16. Tidal volume, vital capacity, inspiratory reserve volume, expiratory reserve volume, and residual volume are measurements of lung capacity and can reflect lung function.

17. External respiration and internal respiration occur when gases move passively along a pressure gradient.

18. Most oxygen is transported through the blood bound to hemoglobin. Carbon dioxide is transported through the blood bound to hemoglobin or in the form of a bicarbonate ion.

19. Breathing is regulated by the nervous system.

20. Disorders of the respiratory system include asthma, emphysema, bronchitis, pneumonia, tuberculosis, botulism, cancer, and congestive heart failure.

Exercises

Complete the exercises for each section after you have read and studied the section. If you cannot answer some questions, or answer them incorrectly, return to the chapter and review this information. You may find it helpful to work on only one section at a time. When you have completed all sections, take the Chapter Test as an indicator of your mastery of this topic.

10.1 Respiration takes place throughout the body

10.2 The respiratory system consists of upper and lower respiratory tracts

Matching

_____ 1. **ventilation**   a. the smallest airways

_____ 2. **external nose**   b. two folds of connective tissue that extend across the larynx

_____ 3. **nasal cavity**   c. tiny air-filled sacs in the lungs that carry out gas exchange

_____ 4. **pharynx**   d. organs of gas exchange

_____ 5. **larynx**   e. two layers of epithelial membrane enclosing each lung

_____ 6. **epiglottis**   f. the movement of air into and out of the lungs

_____ 7. **vocal cords**   g. airways that branch from the trachea and enter the lungs

_____ 8. **glottis**   h. the visible portion of the nose

_____ 9. **trachea**   i. a fluid-filled space between the pleural membranes

_____ 10. **bronchi**   j. the opening into the larynx

_____ 11. **bronchioles**   k. the internal portion of the nose

_____ 12. **lungs**   l. the voice box

_____ 13. **pleural membranes** m. the "windpipe" extending from the larynx to the bronchi

_____ 14. **pleural cavity**   n. the throat

_____ 15. **alveoli**   o. a flexible cartilage flap located at the opening to the larynx

Fill-in-the-Blank

Referenced sections are in parentheses.

16. The _____ system facilitates the exchange of O_2 and CO_2 between the air and the body. (10.1)

17. The exchange of gases between the blood and tissue fluids is _____ respiration. (10.1)

18. The process of using oxygen to produce ATP within cells is _____ respiration. (10.1)

19. In addition to gas exchange, the respiratory system is also responsible for the production of _____. (10.1)

20. The upper respiratory tract consists of the _____ and the _____. (10.2)

continued

21. The lower respiratory tract consists of the _____, _____, _____, and _____. (10.2)

22. The _____ respiratory tract filters, warms, and humidifies air. (10.2)

23. Gas exchange between the air in the alveoli and the blood in the pulmonary capillaries occurs by _____. (10.2)

24. Both food and air pass through the _____. (10.2)

25. The _____ prevents food or liquid from entering the _____. (10.2)

26. The _____ is composed of C-shaped rings of cartilage held together by _____ _____ and _____. (10.2)

27. The walls of the _____ contain fibrous connective tissue and smooth muscle reinforced with cartilage. (10.2)

28. The _____ are marked by the absence of cartilage. (10.2)

29. Friction in the pleural membrane during lung movement is reduced by the presence of _____ in the _____ _____. (10.2)

30. The right lung has _____ lobes and the left lung has _____ lobes. (10.3)

31. The wall of a single alveolus is _____ layer(s) of _____ _____ cells. (10.2)

32. Surface tension within the alveoli is reduced by _____. (10.2)

33. Gas exchange in the lungs occurs between the _____ and the _____ _____. (10.2)

Labeling

Label each of the structures indicated in Figure 10.1.

34. _____

35. _____

36. _____

37. _____

38. _____

39. _____

40. _____

41. _____

42. _____

43. _____

44. _____

45. _____

46. _____

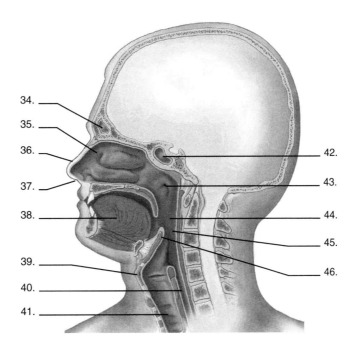

Figure 10.1

10.3 The process of breathing involves a pressure gradient

Fill-in-the-Blank

1. Gas pressure is caused when gas molecules _____.

2. When the volume of a closed space increases, the gas pressure _____.

3. Gases flow from an area of _____ pressure to an area of _____ pressure.

4. Inspiration and expiration occur when air moves down its _____ _____.

Short Answer

Refer to Figure 10.2 to answer the questions below.

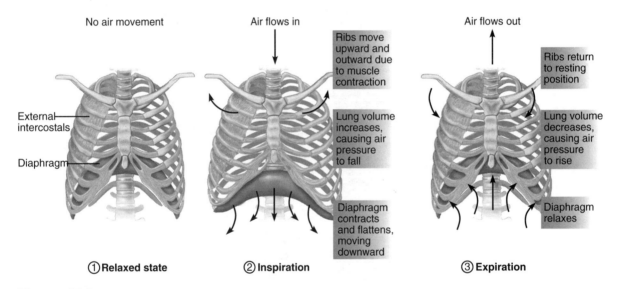

Figure 10.2

5. What muscles are involved in moving air into and out of the lungs?

6. What muscle activity is responsible for inspiration?

7. During inspiration, what change occurs in the volume of the lungs?

8. During inspiration, what change occurs in the air pressure in the lungs?

9. Why does air rush into the lungs during inspiration?

10. As the muscles of respiration relax, what change occurs in lung volume?

11. As the muscles of respiration relax, what change occurs in air pressure in the lungs?

12. Why does air rush out of the lungs during expiration?

Matching

_____ 13. **tidal volume**   a. the amount of additional air that can be inhaled beyond the tidal volume

_____ 14. **vital capacity**   b. the amount of air remaining in the lungs after forcefully exhaling

_____ 15. **inspiratory reserve volume**   c. the maximal volume that can be exhaled after a maximal inhalation

_____ 16. **expiratory reserve volume**   d. the amount of air that can be forcibly exhaled beyond the tidal volume

_____ 17. **residual volume**   e. the amount of air in each normal breath

10.4 Gas exchange and transport occurs passively

Short Answer

Figure 10.3 depicts the partial pressure of oxygen and carbon dioxide in various areas of the body. Differences in partial pressure of a gas create a pressure gradient that drives the movement of the gas. Refer to Figure 10.3 to answer the following questions.

Figure 10.3

1. Gas exchange between the alveoli and the pulmonary capillaries:
   _____ a. What is the PO_2 in the alveoli?
   _____ b. What is the PO_2 in the pulmonary capillaries?
   _____ c. What direction will O_2 move?
   _____ d. What is the PCO_2 in the alveoli?
   _____ e. What is the PCO_2 in the pulmonary capillaries?
   _____ f. What direction will CO_2 move?
   _____ g. What type of respiration is this?

2. Gas exchange between the capillary blood and the interstitial fluid:
   _____ a. What is the PO_2 in the capillary blood?
   _____ b. What is the PO_2 in a cell?
   _____ c. What direction will O_2 move?
   _____ d. What is the PCO_2 in the capillary blood?
   _____ e. What is the PCO_2 in a cell?
   _____ f. What direction will CO_2 move?
   _____ g. What type of respiration is this?

Fill-in-the-Blank

Each formula below represents a process involved in the transport of O_2 or CO_2 in the blood. Answer the questions about each process below the formula for that process.

3. **Hb + O_2 → HbO_2**

   a. HbO_2 is a molecule of _____ bound to _____ and is called _____.

   b. This reaction is _____.

   c. List three conditions that promote the binding of O_2 to hemoglobin: _____, _____, and _____.

4. **Hb + CO_2 → $HbCO_2$**

   a. $HbCO_2$ is a molecule of _____ bound to _____ _____ and is called _____.

   b. What percentage of carbon dioxide is transported in the blood this way? _____

5. $CO_2 + H_2O \rightarrow H_2CO_3 \rightarrow HCO_3 + H^+$

   a. In this reaction, CO_2 combines with water in the blood to produce a _____ ion.

   b. What percentage of carbon dioxide is transported in the blood this way? _____

10.5 The nervous system regulates breathing

10.6 Disorders of the respiratory system

Matching

_____ 1. **respiratory center**   a. viral or bacterial infection resulting in inflammation of the lungs

_____ 2. **stretch receptors**   b. receptors in the arteries that respond to the PO_2 of the blood

_____ 3. **cerebrospinal fluid**   c. uncontrolled growth of abnormal cells that interferes with normal lung function

_____ 4. **carotid and aortic bodies**   d. spasmodic contractions of the bronchi

_____ 5. **asthma**   e. receptors in the lungs that monitor the degree of lung inflation

_____ 6. **emphysema**   f. a bacterial infection that may lead to paralysis of respiratory muscles

_____ 7. **bronchitis**   g. an area of the brain that regulates the rate of breathing

_____ 8. **pneumonia**   h. a condition that leads to accumulation of fluid in the alveoli

_____ 9. **tuberculosis**   i. changes in the H^+ concentration here are detected by cells in the brain

_____ 10. **botulism**   j. a bacterial infection that scars the lungs

_____ 11. **lung cancer**   k. permanent damage to the airways and destruction of the alveoli

_____ 12. **congestive heart failure**   l. an inherited genetic disorder

_____ 13. **cystic fibrosis**   m. inflammation of the bronchi

Chapter Test

Multiple Choice

1. The exchange of gases between inhaled air and the blood is called:
   a. ventilation.
   b. internal respiration.
   c. external respiration.
   d. cellular respiration.

2. The larynx, trachea, bronchi, and lungs make up the:
   a. respiratory system.
   b. upper respiratory tract.
   c. lower respiratory tract.
   d. unrelated respiratory organs.

3. Mucus in the nasal cavity functions to:
   a. maintain the required pH of nasal secretions.
   b. trap dust and pathogens in incoming air.
   c. guide airflow in the proper direction.
   d. enhance the production of cilia.

4. The functions of the larynx include all of the following except:
   a. maintenance of an open airway.
   b. gas exchange.
   c. channeling food and air into the correct area.
   d. assisting in the production of sound.

5. Which of the following structures that influence vocalization are incorrectly matched with their function?
   a. tongue and soft palate—create recognizable sounds
   b. pharynx and nose—act as resonating chambers to amplify and enhance vocal tone
   c. vocal cords—vibrate to produce sound
   d. nasal sinuses—push air through the vocal cords for sound production

6. The incomplete cartilage rings of the trachea allow the trachea to:
   a. maintain a constant diameter at all times.
   b. push air downward with muscle movements.
   c. close when needed to prevent choking.
   d. change diameter to accommodate coughing or heavy breathing.

7. Which of the following represents the correct sequence of airflow?
   a. pharynx, larynx, trachea, bronchi, bronchioles
   b. larynx, pharynx, trachea, bronchi, bronchioles
   c. pharynx, larynx, trachea, bronchioles, bronchi
   d. larynx, pharynx, bronchi, trachea, bronchioles

8. The alveoli:
   a. consist of a single layer of squamous epithelial cells.
   b. are reinforced by cartilage tissue.
   c. do not participate in gas exchange.
   d. fuse to form the bronchioles.

9. The lungs are enclosed by a double pleural membrane. These membranes are attached to the _____ and the _____.
   a. thoracic cavity, pleural cavity
   b. pleural cavity, lung surface
   c. thoracic cavity, lung surface
   d. thoracic cavity, abdominal cavity

10. Gas exchange between inhaled air and the blood is accomplished by the _____ and the _____.
    a. alveoli, pulmonary veins
    b. bronchioles, pulmonary capillaries
    c. alveoli, pulmonary capillaries
    d. alveoli, bronchioles

11. Gases flow passively from an area of _____ pressure to an area of _____ pressure.
    a. high, low
    b. low, high
    c. high, high
    d. low, low

12. Air moves into the lungs when the air pressure in the lungs is _____ the air pressure in the atmosphere.
    a. lower than
    b. higher than
    c. the same as
    d. differences in air pressure are not responsible for airflow

13. Expiration is:
    a. always a passive process.
    b. always an active process.
    c. usually passive, but it can be active.
    d. usually active, but it can be passive.

14. The maximal volume of air that can be exhaled after a maximal inhalation is the:
    a. tidal volume.
    b. vital capacity.
    c. inspiratory reserve volume.
    d. expiratory reserve volume.

15. Measurements of lung capacity are performed with a device called a(n):
    a. volumetric analyzer.
    b. spirometer.
    c. inhalizer.
    d. airflow compression chamber.

16. Under which of the following conditions would oxygen flow from the alveoli into the blood?
    a. PO_2 in the alveoli is 40, PO_2 in the blood is 100
    b. PO_2 in the alveoli is 104, PO_2 in the blood is 104
    c. PCO_2 in the alveoli is 40, PO_2 in the alveoli is 46
    d. PO_2 in the alveoli is 104, PO_2 in the blood is 40

17. In a mixture of gases, the pressure exerted by each gas is called the:
    a. partial pressure.
    b. primary pressure.
    c. residual pressure.
    d. fractional pressure.

18. The flow of CO_2 from deoxygenated blood into the lungs is:
    a. passive.
    b. active.
    c. facilitated.
    d. sometimes passive and sometimes active.

19. Hemoglobin in the blood transports:
    a. oxygen only.
    b. carbon dioxide only.
    c. oxygen and carbon dioxide in equal amounts.
    d. more oxygen than carbon dioxide.
    e. more carbon dioxide than oxygen.

20. Most of the carbon dioxide transported in the blood to the lungs will be transported:
    a. by hemoglobin.
    b. by red blood cells.
    c. as bicarbonate ions in the red blood cells.
    d. as bicarbonate ions in the plasma.

21. A low pH reduces the binding affinity of hemoglobin for oxygen, increasing the release of oxygen and its delivery to the tissues. What produces a condition of low pH in the red blood cells?
    a. conversion of CO_2 to bicarbonate ions
    b. conversion of O_2 to oxyhemoglobin
    c. conversion of CO_2 to carbaminohemoglobin
    d. conversion of bicarbonate ions to CO_2

22. The respiratory center:
    a. monitors the level of oxygen in the blood.
    b. contains stretch receptors that limit lung expansion through negative feedback.
    c. controls the rate of breathing.
    d. is the site of gas exchange between the air and the blood.

23. The concentration of H^+ in the cerebrospinal fluid rises when:
    a. the PO_2 of the arterial blood rises.
    b. the PO_2 of the arterial blood falls.
    c. the PCO_2 of the arterial blood rises.
    d. the PCO_2 of the arterial blood falls.

24. The carotid and aortic bodies are arterial receptors that respond to:
    a. a decrease in the PO_2 of arterial blood.
    b. an increase in the PO_2 of arterial blood.
    c. a decrease in the PO_2 of venous blood.
    d. an increase in the PO_2 of venous blood.

25. A disorder of the respiratory system involving permanent damage to the alveoli is:
    a. asthma.
    b. bronchitis.
    c. emphysema.
    d. pneumonia.

Key Concept Review Questions

Each of the Key Concepts listed at the beginning of this chapter has been rewritten as a question below. After successfully completing the study guide exercises and the Chapter Test, you should be able to answer each of these questions. Refer to the Key Concepts list at the beginning of this chapter to check your answers.

1. What are the four processes encompassed by the term respiration?

2. How would you describe what occurs in each of the four processes listed above?

3. What are the components and function of the upper respiratory tract?

4. What are the components and function of the lower respiratory tract?

5. What acts as a common passageway for food, liquid, and air? What prevents food and liquid from moving into the airways?

6. Where and how is sound produced?

7. How would you compare the amount of cartilage present in the trachea, bronchi, and bronchioles?

8. What structures are located at the end of the bronchioles? What process occurs here?

9. What body cavity contains the lungs? What membrane encloses the lungs?

10. What are two skeletal muscles that contribute to lung function?

11. What happens to the air pressure in the lungs as the lung volume increases?

12. How would you compare the air pressure in the atmosphere with air pressure in the lungs when air flows into the lungs?

13. What is inspiration? Why is it considered an active process?

14. What muscle activity reduces lung volume? Compare the air pressure in the atmosphere with the air pressure in the lungs when air flows out of the lungs.

15. What is expiration? Is it active or passive?

16. What are five measurements of lung capacity?

17. Do gases move with or against a pressure gradient during external and internal respiration?

18. How is most oxygen transported in the blood? How is carbon dioxide transported in the blood?

19. What body system regulates breathing?

20. What are eight disorders of the respiratory system discussed in the chapter?

The Nervous System— Integration and Control

Chapter Summary and Key Concepts

After reading and studying this chapter you should know the following:

Sections 11.1, 11.2, 11.3, 11.4

1. The three characteristics of the nervous system are (1) it requires information, (2) it integrates information, and (3) it can respond to information very quickly.

2. The two principal parts of the nervous system are the central nervous system (CNS), consisting of the brain and spinal cord, and the peripheral nervous system (PNS), consisting of all nervous tissue not in the CNS.

3. Neurons are nervous system cells specialized to receive and transmit information.

4. The three classifications of neurons are (1) sensory neurons that carry information to the CNS, (2) motor neurons that carry information away from the CNS, and (3) interneurons that communicate between nervous system components.

5. The parts of a neuron are the cell body, dendrites, axon hillock, axon, myelin sheath, nodes of Ranvier, axon terminals, and axon bulbs.

6. The slight difference between the voltage inside and outside a cell is the resting membrane potential.

7. A graded potential occurs when chemically sensitive channels allow ion movements that depolarize or hyperpolarize the membrane.

8. An action potential is a reversal in the voltage difference across a membrane and is triggered when the sum of all graded potentials reaches a threshold level.

9. The three events of an action potential are depolarization, repolarization, and reestablishment of the resting potential.

10. During an action potential, depolarization occurs when sodium moves into the cell, repolarization occurs when potassium moves out of the cell, and reestablishment of the resting membrane potential is accomplished by the sodium-potassium pump.

11. Neurotransmitters are chemicals that carry information across a synapse from one neuron to a target cell. Neurotransmitters that are excitatory depolarize the postsynaptic cell, and neurotransmitters that are inhibitory hyperpolarize the postsynaptic cell.

12. Postsynaptic cells integrate and process information.

Sections 11.5, 11.6, 11.7, 11.8, 11.9, 11.10, 11.11, 11.12

13. A nerve contains many axons bundled together. Cranial nerves attach to the brain, and spinal nerves attach to the spinal cord.

14. The PNS is divided into the sensory nervous system that carries information to the CNS, and the motor nervous system that carries information away from the CNS.

15. The motor nervous system branches into the somatic division that controls skeletal muscle, and the autonomic division that controls smooth muscle, cardiac muscle, and glands. The autonomic division branches further into the sympathetic and parasympathetic nervous systems.

16. The three regions of the brain are the hindbrain that includes the medulla oblongata, pons, and cerebellum; the forebrain that includes the thalamus, hypothalamus, limbic system, and cerebrum; and the midbrain.

17. The cerebrum is responsible for higher functions and is divided into the occipital, temporal, parietal, and frontal lobes.

18. The reticular activating system is responsible for levels of sleep and wakefulness. The limbic system in the forebrain is the site of emotion and basic behaviors.

19. The limbic system manages short-term memories for a few hours. Long-term memories occur in the cerebrum and may last for years.

20. Disorders of the nervous system include trauma, infections, abnormal neural activity, and tumors.

Exercises

Complete the exercises for each section after you have read and studied the section. If you cannot answer some questions, or answer them incorrectly, return to the chapter and review this information. You may find it helpful to work on only one section at a time. When you have completed all sections, take the Chapter Test as an indicator of your mastery of this topic.

11.1 The nervous system has two principal parts

11.2 Neurons are the communication cells of the nervous system

11.3 Neurons initiate action potentials

11.4 Information is transferred from a neuron to its target

Matching

___ 1. **central nervous system**
___ 2. **peripheral nervous system**
___ 3. **sensory division**
___ 4. **motor division**
___ 5. **neurons**
___ 6. **sensory neurons**
___ 7. **interneurons**
___ 8. **motor neurons**
___ 9. **axon hillock**
___ 10. **neuroglial cell**
___ 11. **resting potential**
___ 12. **graded potential**
___ 13. **summation**
___ 14. **threshold**
___ 15. **action potential**
___ 16. **synapse**
___ 17. **neurotransmitter**

a. cells that transmit information between components of the central nervous system
b. cells of the nervous system specialized for communication
c. a membrane voltage that triggers an action potential
d. a division of the nervous system consisting of the brain and spinal cord
e. a chemical substance that transmits a signal from a neuron to its target
f. a transient local change in the resting potential of a cell
g. the division of the nervous system containing all components that lie outside the brain and spinal cord
h. a slight difference in voltage that exists between the inside and the outside of a cell
i. a sudden, temporary reversal of voltage difference across the cell membrane
j. the division of the nervous system that carries information away from the brain and spinal cord
k. area of the cell body from which an axon originates
l. a specialized junction between two cells
m. neurons that carry information from the CNS to the body
n. the division of the nervous system that carries information to the brain and spinal cord
o. provides support and protection to neurons
p. the ability of a neuron to respond to the total effect of many incoming signals
q. neurons that respond to a stimulus and transmit information about the stimulus to the CNS

Fill-in-the-Blank

Referenced sections are in parentheses

18. The myelin sheaths of neurons in the PNS function to save _____, speed _____, and repair damaged _____. (11.2)

19. Neurons of the CNS are unable to repair injured axons because the myelin sheath is produced by _____ and degenerates when injured. (11.2)

20. The _____ pump is important in maintaining the resting membrane potential of cells. (11.2)

21. The resting membrane potential of a neuron is about _____, and the inside of the cell is _____ compared to the outside. (11.2)

22. An action potential occurs because the axon contains _____ sensitive ion channels. (11.3)

23. The sodium-potassium pump actively transports _____ Na^+ out of the cell, and _____ K^+ into the cell. (11.3)

24. All action potentials are equal in form and magnitude; this is referred to as the _____ phenomenon. (11.3)

25. Acetylcholine is a common neurotransmitter that has an _____ effect on skeletal muscle. (11.4)

26. The _____ _____ is the cell membrane of the neuron that is sending information at a synapse. (11.4)

27. The _____ _____ is the cell membrane of the cell that receives information at a synapse. (11.3)

28. Neurotransmitters that hyperpolarize a postsynaptic cell are _____. (11.4)

29. In _____ one neuron receives input from many neurons. (11.4)

Labeling

Label the indicated areas of Figure 11.1 with a structural term and a function.

Structure
a. axon
b. dendrite
c. cell body
d. axon terminal
e. axon bulb
f. myelin sheath
g. Schwann cell
h. node of Ranvier

Function
i. produces myelin
j. transmits information in the form of an electrical impulse
k. contains neurotransmitters in membrane-bound vesicles
l. a terminal branch of the axon
m. receives incoming information
n. a short uninsulated gap in the axon
o. the main body of the neuron
p. an insulating layer surrounding the axon that speeds the rate of conduction

30. ____ , ____
31. ____ , ____
32. ____ , ____
33. ____ , ____
34. ____ , ____
35. ____ , ____
36. ____ , ____
37. ____ , ____

Figure 11.1

Completion

Complete the table below as you refer to Figure 11.2. The first row has been completed as an example.

Figure 11.2

Name of Event	Region of Figure 11.2	Description of Electrical Activity
38. resting membrane potential	39. a	a difference in voltage between the inside and outside of a cell, usually about –70 mV
40.	41.	a graded potential that makes the membrane voltage more negative
42.	43.	potassium ions move out of the axon as potassium channels open
44.	45.	electrical event in the axon— a reversal of the voltage difference across the membrane
46.	47.	voltage level achieved by summation that triggers an action potential

continued

Name of Event	Region of Figure 11.2	Description of Electrical Activity
48.	49.	a graded potential that makes the membrane voltage less negative
50.	51.	sodium ions move into the axon as sodium channels open
52.	53.	the cumulative effect of many graded potentials

Word Choice

Circle the term or phrase that correctly completes each sentence.

54. Graded potentials (are/are not) self-propagating.

55. The opening of sodium channels on the axon allows Na$^+$ to move (into/out of) the cell.

56. An action potential (is/is not) self-propagating.

57. The speed of an action potential increases as the diameter of the axon (increases/decreases).

58. When an action potential arrives at the axon bulb, calcium channels (open/close) and Ca^{++} moves (into/out of) the bulb.

59. The effect of a neurotransmitter is relatively (short-lived/long-lived).

60. Muscle cells (do/do not) process and integrate information because they receive input from (one/multiple) neuron(s).

61. The amount of K$^+$ that leaks out of a cell is (greater/less) than the amount of Na$^+$ that leaks into the cell.

11.5 The PNS relays information between tissues and the CNS

11.6 The brain and spinal cord comprise the CNS

Crossword Puzzle

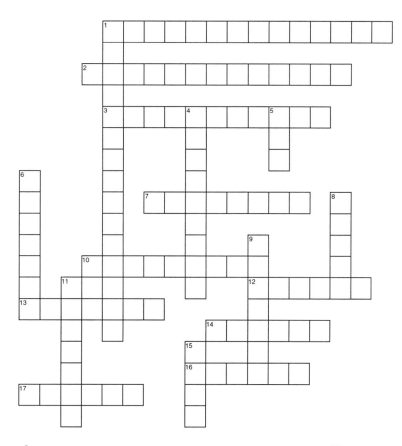

Across

1. The cell bodies of _____ neurons lie outside the CNS
2. The cell bodies of _____ neurons lie in the CNS
3. The branch of the autonomic nervous system that prepares the body for "fight or flight"
7. Connective tissue membranes that surround the CNS
10. Cerebrospinal fluid is secreted into the _____ of the brain
12. Bundles of many axons wrapped in a connective sheath
13. _____ nerves that connect directly with the brain
14. The root of a spinal nerve that contains sensory neurons
16. A spinal _____ is an involuntary response that bypasses the brain
17. _____ nerves connect with the spinal cord

Down

1. The branch of the automatic nervous system that predominates during relaxation
4. The branch of the PNS that controls smooth and cardiac muscle
5. Pathways of the autonomic nervous system require _____ neurons
6. The branch of the PNS that controls skeletal muscle
8. The _____ matter of the CNS contains nerve tracts
9. The root of a spinal nerve that contains motor neurons
11. A collection of cell bodies of postsynaptic neurons
15. The _____ matter of the CNS contains unmyelinated axons and cell bodies

11.7 The brain processes and acts on information

11.8 Brain activity continues during sleep

11.9 The limbic system is the site of emotion and basic behaviors

11.10 Memory involves storing and retrieving information

Matching

_____ 1. **occipital lobe**
_____ 2. **hindbrain**
_____ 3. **temporal lobe**
_____ 4. **parietal lobe**
_____ 5. **frontal lobe**
_____ 6. **reticular activating system**
_____ 7. **limbic system**
_____ 8. **short-term memory**
_____ 9. **long-term memory**

a. neurons that transmit action potentials to the cerebrum to maintain alertness
b. cerebral area that interprets auditory information and comprehends spoken and written language
c. a group of neuronal pathways in the thalamus, hypothalamus, and cerebrum associated with emotions and memory
d. the oldest and most primitive part of the brain
e. memory stored by a permanent change to neurons in the cerebral cortex
f. cerebral area that processes visual information
g. cerebral area that receives and interprets sensory information
h. memory that occurs in the limbic system
i. cerebral area that initiates motor activities and is responsible for conscious thought

Labeling

Label the indicated areas in Figure 11.3 with a structural term and a function.

Structure	Function
a. pons	g. controls automatic functions of the internal organs
b. cerebrum	h. receives, processes, and transfers information
c. cerebellum	i. aids the medulla in regulating respiration
d. thalamus	j. bridges the two cerebral hemispheres

Structure	Function
e. medulla oblongata	k. involved in decision making and conscious thought
f. corpus callosum	l. coordinates basic subconscious body movements

Figure 11.3

11.11 Psychoactive drugs affect higher brain functions

11.12 Disorders of the nervous system

Short Answer

1. What is a psychoactive drug?

2. Why can psychoactive drugs affect the brain when many other drugs cannot?

3. What is the common mode of action for psychoactive drugs?

4. Why does tolerance develop?

5. Define "addiction."

Completion

Name each disorder of the nervous system described in the table.

Disorder	Description
6.	blow to the head or neck with loss of consciousness
7.	violent trauma to the vertebrae that compresses, tears, or severs the spinal cord
8.	an infection resulting in inflammation of the brain
9.	an infection resulting in inflammation of the meninges
10.	viral brain infection transmitted to humans by direct contact with infected animals, often through a bite
11.	recurrent episodes of abnormal electrical activity in the brain, often of unknown causes
12.	progressive degenerative disorder caused by the loss of dopamine-releasing neurons
13.	changes in structure and activity of frontal and temporal lobes leading to mental impairment
14.	abnormal growth in or on the brain

Chapter Test

Multiple Choice

1. The functional divisions of the peripheral nervous system are:
   a. sympathetic and parasympathetic.
   b. sensory and motor.
   c. autonomic and somatic.
   d. autonomic and sensory.

2. Neurons of the PNS that transmit information to the CNS are:
   a. sensory neurons.
   b. motor neurons.
   c. interneurons.
   d. sympathetic neurons.

3. Axons of sensory neurons originate from the:
   a. cell body.
   b. axon hillock.
   c. dendrite.
   d. axon bulb.

4. The myelin sheath of PNS neurons is produced by _____, while the myelin sheath of CNS neurons is produced by _____.
   a. oligodendrocytes, Schwann cells
   b. neuroglia, oligodendrocytes
   c. Schwann cells, oligodendrocytes
   d. Schwann cells, Schwann cells

5. Which of the following is *not* a function of the myelin sheath?
   a. saving the neuron energy
   b. speeding the transmission of impulses
   c. regenerating damaged neurons in the PNS
   d. initiating electrical impulses on the axon

6. Which of the following characteristics of the neuron cell membrane contribute to the maintenance of the resting membrane potential?
   a. The membrane is more permeable to sodium than to potassium.
   b. Sodium leaks out of the membrane more easily than potassium.
   c. The sodium-potassium pump transports three sodium ions out of the cell for every two potassium ions it pumps into the cell.
   d. a and c
   e. all of the above

7. Graded potentials:
   a. are permanent changes in the resting potential of a membrane.
   b. occur in a local area of the membrane.
   c. always depolarize the membrane.
   d. grow stronger as they spread through the cell.

8. Summation may occur:
   a. when rapid stimulation of a neuron causes the membrane voltage to reach threshold levels.
   b. when many neurons all cause hyperpolarizing graded potentials.
   c. when the presence of the axon hillock lowers the normal threshold level.
   d. when the cell body contains both chemically sensitive and voltage-sensitive channels.

9. Repolarization occurs when:
   a. voltage-sensitive calcium channels open.
   b. chemically sensitive sodium channels close.
   c. potassium channels open.
   d. sodium leaks out of the cell.

10. Which of the following would occur in a neuron if calcium could not enter the axon bulb?
    a. absence of graded potentials
    b. absence of action potentials
    c. inability to manufacture neurotransmitters
    d. inability to release neurotransmitters

11. The absolute refractory period:
    a. occurs during transmission of an action potential.
    b. occurs shortly after transmission of an action potential.
    c. occurs shortly before transmission of an action potential.
    d. reflects the actions of the sodium-potassium pump.

12. Action potentials are "all-or-none" events. Which of the following statements contradicts this?
    a. Besides generating an action potential, the only additional information a neuron can transmit is the frequency of stimulation.
    b. If a threshold level of stimulation is reached, the action potential will always be of the same form and strength.
    c. Summation in a postsynaptic neuron can occur when the frequency of stimulation increases.
    d. The level of neurotransmitter bound to the postsynaptic membrane affects the speed of transmission of the action potential.

13. Excitatory neurotransmitters:
    a. depolarize the postsynaptic membrane.
    b. hyperpolarize the postsynaptic membrane.
    c. always initiate an action potential.
    d. can depolarize or hyperpolarize the postsynaptic membrane depending on the type of receptors present.

14. Which of the following must occur immediately before neurotransmitter can be released into the synaptic cleft?
    a. an action potential
    b. a graded potential
    c. reuptake of neurotransmitter by the presynaptic membrane
    d. opening of calcium channels on the presynaptic membrane

15. When one neuron receives input from many neurons, this is called:
    a. summation.
    b. convergence.
    c. divergence.
    d. multiple stimulation.

16. The ventral root of a spinal nerve contains:
    a. sensory neurons.
    b. motor neurons.
    c. sensory and motor neurons.
    d. cell bodies only.

17. Motor pathways requiring a preganglionic and a postganglionic neuron belong to the _____ nervous system.
    a. somatic division of the
    b. sensory division of the
    c. central
    d. autonomic

18. If your heart rate and blood pressure increase in response to a crisis, you are experiencing the immediate effects of the _____ nervous system.
    a. sensory
    b. sympathetic
    c. parasympathetic
    d. reflexive

19. Which of the following is *not* one of the meninges of the central nervous system?
    a. dura mater
    b. pia mater
    c. ventral
    d. arachnoid

20. White matter of the spinal cord contains _____,
    while gray matter contains _____.
    a. nerve tracts, cell bodies
    b. cell bodies, nerves
    c. unmyelinated axons, myelinated axons
    d. motor neurons, sensory neurons

21. Which part of the brain receives, processes, and transfers information?
    a. cerebellum
    b. medulla oblongata
    c. corpus callosum
    d. thalamus

22. Higher brain function occurs in the:
    a. cerebrum.
    b. cerebellum.
    c. hindbrain.
    d. limbic system.

23. In the primary somatosensory area:
    a. the brain responds to visual input.
    b. conscious thought is coordinated with motor responses.
    c. body parts that are extremely sensitive are represented by a greater area than less sensitive body parts.
    d. body parts that are coordinated with motor responses are represented by a greater area than body parts that are not.

24. During which stage of sleep do skeletal muscles relax and eye and body movements cease?
    a. stage 1
    b. stage 2
    c. stage 3
    d. stage 4
    e. stage 5

25. The site of emotions and basic behavior is the:
    a. parietal lobe of the cerebral cortex.
    b. thalamus.
    c. limbic system.
    d. hindbrain.

Key Concept Review Questions

Each of the Key Concepts listed at the beginning of this chapter has been rewritten as a question below. After successfully completing the study guide exercises and the Chapter Test, you should be able to answer each of these questions. Refer to the Key Concepts list at the beginning of this chapter to check your answers.

1. What are the three characteristics of the nervous system?
2. What are the two principal parts of the nervous system and what components does each contain?
3. What cells of the nervous system are specialized to receive and transmit information?
4. What are the three classifications of neurons?
5. What are the parts of a neuron?
6. What is the resting membrane potential?
7. What causes a graded potential to occur?
8. What is an action potential?
9. What are the three events of an action potential?
10. During an action potential, what ion activity causes each of the three events?
11. What is a neurotransmitter? How do excitatory neurotransmitters differ from inhibitory neurotransmitters in their effects on the postsynaptic cell membrane?
12. What synaptic cells integrate and process information?
13. What is a nerve? How do spinal nerves differ from cranial nerves?
14. What are the two divisions of the peripheral nervous system? Which division carries information to the brain? Which division carries information away from the brain?
15. What are the branches of the motor nervous system? What are the branches of the autonomic division?
16. What are the regions of the brain?
17. What is the general function of the cerebrum? List the lobes of the cerebrum.
18. What is the function of the reticular activating system? What is the function of the limbic system?
19. How does short-term memory differ from long-term memory in location and duration?
20. What are the four general categories of nervous system disorders discussed in this chapter?

12

Sensory Mechanisms

Chapter Summary and Key Concepts

After reading and studying this chapter you should know the following:

Sections 12.1, 12.2, 12.3

1. Receptors convert the energy of a stimulus into impulses in sensory neurons.

2. Receptors are classified according to the type of stimulus energy they convert.

3. The brain interprets nerve impulses based on their origin and frequency.

4. Adaptation occurs in some receptors and is the cessation of impulses in the presence of a continued stimulus.

5. The somatic senses are temperature, touch, vibration, pressure, pain, and awareness of body position and movements. The somatic senses arise from receptors located throughout the body.

6. The special senses are taste, smell, hearing, balance, and vision. The special senses arise from receptors in specialized areas of the body.

7. Taste depends on chemoreceptors located in taste buds on the tongue and mouth.

8. Smell depends on chemoreceptors located in the nasal passages.

Sections 12.4, 12.5

9. Mechanoreceptors in the ear detect sound waves.

10. The ear consists of an outer ear, middle ear, and inner ear.

11. The outer ear is called the pinna and channels sound waves.

12. The middle ear consists of three small bones that amplify sound and transmit the waves to the inner ear.

13. The inner ear converts sound to impulses in the cochlea and transmits information to the brain via the auditory nerve.

14. The inner ear also contains the semicircular canals that sense rotational movement, and the vestibular apparatus that senses static head position and linear acceleration.

Sections 12.6, 12.7

15. The structures of the eye convert visual stimuli to impulses that are transmitted to the brain via the optic nerve.

16. Light entering the eye is bent by the cornea and the lens to focus an image.

17. Accommodation is the adjustment of lens curvature that allows the eye to focus on objects at any distance.

18. The photoreceptor cells of the eye are the photopigment-containing rods and cones.

19. Rods function in dim light, and cones respond to color.

20. Disorders of sensory mechanisms include retinal detachment, glaucoma, cataracts, color blindness, otitis media, and Ménière's syndrome.

Exercises

Complete the exercises for each section after you have read and studied the section. If you cannot answer some questions, or answer them incorrectly, return to the chapter and review this information again. You may find it helpful to work on only one section at a time. When you have completed all sections, take the Chapter Test as an indicator of your mastery of this topic.

12.1 Receptors receive and convert stimuli

12.2 Somatic sensations arise from receptors throughout the body

12.3 Taste and smell depend on chemoreceptors

Matching

____ 1. **stimulus**　　　　　a. receptors that respond to forms of mechanical energy

____ 2. **receptor**　　　　　b. receptors that respond to light

____ 3. **sensation**　　　　　c. sensory neurons located in the nasal passages

____ 4. **perception**　　　　d. mechanoreceptors that monitor muscle length

____ 5. **mechanoreceptors**　e. sensory input that causes a change within or outside the body

____ 6. **thermoreceptors**　　f. a cluster of taste cells and support cells on the tongue and mouth

continued

_____ 7. **pain receptors** g. sensations of temperature, touch, vibration, pressure, and pain

_____ 8. **chemoreceptors** h. a structure specialized to receive a certain stimulus

_____ 9. **photoreceptors** i. sensory neurons stop sending impulses while the stimulus is still present

_____ 10. **receptor adaptation** j. conscious awareness of a stimulus

_____ 11. **somatic sensation** k. interpretation of a sensation

_____ 12. **special senses** l. receptors that respond to the presence of chemicals

_____ 13. **muscle spindles** m. receptors that respond to heat or cold

_____ 14. **taste buds** n. sensations of taste, smell, hearing, balance, and vision

_____ 15. **olfactory receptor cells** o. receptors that respond to tissue damage or excessive pressure or temperature

Word Choice

Circle the term or phrase that correctly completes each sentence.

16. Individuals are (conscious/not conscious) of receptors that function in negative feedback loops to maintain homeostasis.

17. Receptors in the skin for light touch and pressure adapt (slowly/quickly).

18. Body parts with the greatest sensory sensitivity involve the (fewest/most) neurons.

19. Receptors for (slow pain/fast pain) are usually located near the surface of the body.

Fill-in-the-Blank

Referenced sections are in parentheses.

20. Taste buds convert a chemical signal into a(n) _____ _____. (12.3)

21. The four major categories of taste are _____, _____, _____, and _____. (12.3)

22. Taste buds contain chemoreceptors for _____, while olfactory receptor cells have chemoreceptors for _____. (12.3)

23. The receptors of olfactory cells are modified dendritic endings of sensory neurons and are called _____. (12.3)

Chapter 12 *Sensory Mechanisms* **161**

12.4 Hearing: Mechanoreceptors detect sound waves

12.5 The inner ear plays an essential role in balance

Labeling

Use the terms to label Figure 12.1.

middle ear
auditory canal
outer ear
incus
vestibule
round window

pinna
semicircular canals
cochlear nerve
cochlea
inner ear
eustachian tube

malleus
oval window
vestibular nerve
tympanic membrane
stapes

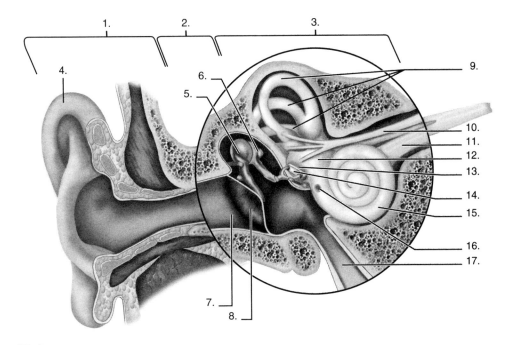

Figure 12.1

1. _____
2. _____
3. _____
4. _____
5. _____
6. _____
7. _____
8. _____
9. _____
10. _____
11. _____
12. _____
13. _____
14. _____
15. _____
16. _____
17. _____

Paragraph Completion

Use the terms to complete the paragraph, and then repeat this exercise with the terms covered.

amplitude	eardrum	neurotransmitter
pinna	malleus	vestibular
basilar	stapes	cochlea
waves	oval window	auditory
tectorial	incus	frequency
base	round window	tip

Sounds are (18) _____ of compressed air. The loudness of a sound depends on the (19) _____ of the waves; the tone of a sound depends on the (20) _____ of the waves. Sound waves first arrive at the (21) _____ and cause vibrations of the (22) _____. These vibrations are passed on to the (23) _____, (24) _____, and (25) _____ of the middle ear, which amplify the sound and lead to vibration of a small membrane called the (26) _____. This membrane transmits the vibrations to the fluid in the (27) _____ canal of the (28) _____. Pressure waves in the fluid of the cochlea vibrate the (29) _____ membrane, moving the hair cells that are embedded in the (30) _____ membrane. Bending of the hair cells causes the release of a (31) _____ and generates an impulse in the (32) _____ nerve. High-pitched tones vibrate the basilar membrane closer to the (33) _____ of the cochlea, while low-pitched tones vibrate the membrane closer to the (34) _____ of the cochlea. Fluid vibrations are dissipated when they reach the (35) _____.

Matching

____ 36. **vestibular apparatus**   a. hard crystals embedded in gel in the chambers of the vestibule

____ 37. **semicircular canals**   b. bulging region at the end of each semicircular canal

____ 38. **vestibule**   c. inner ear structure consisting of the vestibule and the semicircular canals

____ 39. **ampulla**   d. senses static head position and linear acceleration

_____ 40. **cupula**   e. consists of the utricle and saccule

_____ 41. **otoliths**   f. three fluid-filled canals that sense rotational movement

_____ 42. **utricle and saccule**   g. a gel-like dome located in the ampulla and embedded with hairs of sensory neurons

12.6 Vision: Detecting and interpreting visual stimuli

12.7 Disorders of sensory mechanisms

Labeling

A. *Use the terms to label Figure 12.2.*

pupil	choroid	iris	aqueous humor
sclera	fovea	retina	Canal of Schlemm
optic disk	lens	cornea	ciliary muscle
optic nerve	vitreous humor		

Figure 12.2

1. _____
2. _____
3. _____
4. _____
5. _____
6. _____
7. _____
8. _____
9. _____
10. _____
11. _____
12. _____
13. _____
14. _____

B. *Refer to Figure 12.2 as you write the function of each structure listed below.*

15. sclera: _____

16. cornea: _____

17. aqeous humor: _____

18. iris: _____

19. lens: _____

20. ciliary muscle: _____

21. vitreous humor: _____

22. retina: _____

Fill-in-the-Blank

Referenced sections are in parentheses.

23. The photoreceptor cells of the retina are called _____ and _____. (12.6)

24. The flattened disks of photoreceptor cells contain a protein called a _____. (12.6)

25. _____ is the photopigment found in rods. (12.6)

26. _____ are photoreceptor cells that respond to color. (12.6)

27. The innermost layer of the retina consists of neurons called _____ _____. (12.6)

Completion

Write the name of each sensory disorder described in the table.

Sensory Disorder	Description
28.	the retina separates from the choroid, often caused by a blow to the head
29.	blockage in the Canal of Schlemm prevents drainage of the aqueous humor, leading to pressure in the eye
30.	a decrease in the normal transparency of the lens
31.	inflammation of the middle ear
32.	a chronic condition of the inner ear, characterized by recurrent episodes of dizziness, nausea, and hearing loss
33.	distant objects are focused in front of the retina

Chapter Test

Multiple Choice

1. Receptors are classified according to:
   a. their location in the body.
   b. the type of cell they arise from.
   c. the type of stimulus energy they convert.
   d. how easily they generate an action potential.

2. Pain receptors adapt slowly or not at all because:
   a. lack of adaptation is important to survival.
   b. only mechanoreceptors are capable of adaptation.
   c. adaptation occurs but is ignored by the brain.
   d. the stimulus is too strong for adaptation to occur.

3. Pacinian corpuscles detect:
   a. light touch.
   b. deep pressure.
   c. temperature.
   d. ongoing pressure.

4. Muscle spindles are:
   a. chemoreceptors.
   b. photoreceptors.
   c. thermoreceptors.
   d. mechanoreceptors.

5. Sensations are categorized as:
   a. somatic or special.
   b. sensory or motor.
   c. adaptive or nonadaptive.
   d. central or peripheral.

6. Chemoreceptors for taste are specific for a chemical called a:
   a. tastant.
   b. papillae.
   c. taste bud.
   d. taste hair.

7. The chemoreceptors for smell are located on:
   a. sensory neurons.
   b. epithelial cells.
   c. naked nerve endings.
   d. glandular cells.

8. The olfactory bulb is:
   a. the site of origin for olfactory receptor cells.
   b. an area of the nose where olfactory cells are highly concentrated.
   c. an area of the brain where olfactory receptor cells synapse with other neurons.
   d. the area of the brain where smells are interpreted.

9. Sounds are waves of compressed air. The loudness of a sound is determined by:
   a. the frequency of the waves.
   b. the amplitude of the waves.
   c. the temperature of the air.
   d. the volume of air.

10. The eardrum is referred to as the:
    a. pinna.
    b. tympanic membrane.
    c. vestibule.
    d. malleus.

11. Which part of the ear generates impulses for sound?
    a. ampulla
    b. utricle
    c. saccule
    d. cochlea

12. Vibrations are passed from the middle ear to the inner ear through the:
    a. cochlear window.
    b. round window.
    c. oval window.
    d. tympanic window.

13. The base of the cochlear duct is formed by the:
    a. tectorial membrane.
    b. vestibular membrane.
    c. cochlear membrane.
    d. basilar membrane.

14. The structure of the inner ear that converts pressure waves to impulses is the:
    a. vestibular canal.
    b. cochlear duct.
    c. organ of Corti.
    d. round window.

15. The mechanoreceptor cells of the ear are:
    a. hair cells.
    b. tectorial cells.
    c. vestibular cells.
    d. tympanic cells.

16. The semicircular canals sense:
    a. rotational movement of the head and body.
    b. inertia.
    c. linear acceleration.
    d. vertical deceleration.

17. Otoliths are found in the:
    a. middle ear.
    b. cochlea.
    c. ampulla.
    d. vestibular apparatus.

18. The photoreceptors of the eye are:
    a. the optic nerve.
    b. the ciliary muscle.
    c. rods and cones.
    d. choroid pigment cells.

19. The first structure of the eye through which light passes is the:
    a. sclera.
    b. choroid.
    c. lens.
    d. cornea.

20. The fluid found between the lens and the cornea is the:
    a. aqueous humor.
    b. vitreous humor.
    c. ciliary fluid.
    d. synovial fluid.

21. Accommodation occurs as a result of the actions of the:
    a. cornea.
    b. ciliary muscle.
    c. vitreous humor.
    d. retina.

22. The layer of bipolar cells in the retina:
    a. receives information from the rods and cones.
    b. receives information from ganglion cells.
    c. absorbs light to stimulate rods and cones.
    d. possesses axons that will become the optic nerve.

23. Rhodopsin is:
    a. an enzyme produced by the ciliary muscle.
    b. a nutrient found in the vitreous humor.
    c. the photopigment found in rods.
    d. a chemoreceptor that monitors blood supply to the retina.

24. The optic disk is the area of the eye:
    a. with the highest concentration of cones.
    b. with the highest concentration of rods.
    c. where the optic nerve exits.
    d. where the retina attaches to the sclera.

25. Glaucoma is a condition in which:
    a. the lens becomes less transparent.
    b. pressure builds up inside the eye.
    c. the retina detaches from the choroid.
    d. the elasticity of the ciliary muscle decreases.

Key Concept Review Questions

Each of the Key Concepts listed at the beginning of this chapter has been rewritten as a question below. After successfully completing the study guide exercises and the Chapter Test, you should be able to answer each of these questions. Refer to the Key Concepts list at the beginning of this chapter to check your answers.

1. What do receptors do with a stimulus?
2. How are receptors classified?
3. What characteristics does the brain use to interpret a nerve impulse?
4. What is adaptation? Does it occur in all receptors?
5. What are the somatic senses? Where are the receptors for somatic senses located?
6. What are the special senses? Where are the receptors for the special senses located?
7. What type of receptor is responsible for taste? Where are these receptors located?

8. What type of receptor is responsible for smell? Where are these receptors located?
9. What type of receptors detect sound waves?
10. What are the three main parts of the ear?
11. What is the outer ear called? What is the function of the outer ear?
12. What structures are found in the middle ear? What is the function of the middle ear?
13. What role does the inner ear play in sound? What general part of the inner ear is responsible for hearing?
14. What part of the inner ear functions in sensing rotational movement? What part of the inner ear functions in sensing head position and linear acceleration?
15. What does the eye do with visual stimuli?
16. What structures of the eye bend light to focus an image?
17. What is accommodation?
18. What are the photoreceptor cells of the eye?
19. What is the specific function of rods and cones?
20. What are six disorders of sensory mechanisms discussed in this chapter?

13

The Endocrine System

Chapter Summary and Key Concepts

After reading and studying this chapter you should know the following:

Sections 13.1, 13.2, 13.3

1. The endocrine system is a collection of specialized cells, tissues, and glands that produce hormones. Hormones are secreted into extracellular space and enter the blood or lymph.

2. Hormones act only on their target cells.

3. Endocrine control of body activities is slower than nervous control.

4. Hormones are classified as steroid or nonsteroid.

5. Steroid hormones are synthesized from cholesterol, are lipid soluble, and exert their effect by binding to intracellular receptors and activating or inhibiting genes.

6. Nonsteroid hormones are synthesized from amino acids, are not lipid soluble, and exert their effect by binding to a cell membrane and activating a second messenger molecule.

7. Many hormones participate in negative feedback loops that maintain homeostasis.

8. The hypothalamus is a region of the brain that interacts with the endocrine system primarily through the pituitary gland.

9. The pituitary gland consists of a posterior lobe and an anterior lobe.

10. The posterior pituitary stores ADH and oxytocin secreted by neuroendocrine cells in the hypothalamus.

11. The anterior pituitary produces ACTH, TSH, FSH, LH, PRL, and GH. Releasing and inhibiting hormones produced by the hypothalamus control the release of these hormones.

Sections 13.4, 13.5, 13.6, 13.7, 13.8

12. The pancreas secretes insulin when blood sugar levels are high, and glucagon when blood sugar levels are low. The pancreas also secretes somatostatin, a hormone that inhibits insulin and glucagon secretion.

13. The adrenal cortex produces cortisol that helps maintain blood glucose levels, and aldosterone that helps maintain water balance. The adrenal cortex also produces small amounts of sex hormones.

14. The adrenal medulla produces epinephrine and norepinephrine. The release of these hormones is stimulated by the sympathetic nervous system, and they contribute to the fight-or-flight response.

15. The thyroid gland produces thyroxin that regulates the production of ATP, and calcitonin that decreases blood calcium levels.

16. The parathyroid glands produce parathyroid hormone that increases the blood calcium levels.

17. The testes produce testosterone and the ovaries produce estrogen and progesterone. These sex hormones regulate sexual development, activities, and secondary sex characteristics.

18. The thymus gland secretes thymosin and thymopoietin for the maturation of T cells; the pineal gland secretes melatonin that may play a role in sleep/wake cycles; and the heart, kidneys, and digestive system all secrete hormones specialized for their functions.

Sections 13.9, 13.10

19. Histamine, prostaglandins, nitric oxide, and growth factors are chemical messengers similar to hormones but with only local effects.

20. Disorders of the endocrine system include hyperthyroidism, hypothyroidism, Addison's disease, and Cushing's syndrome.

Exercises

Complete the exercises for each section after you have read and studied the section. If you cannot answer some questions, or answer them incorrectly, return to the chapter and review this information. You may find it helpful to work on only one section at a time. When you have completed all sections, take the Chapter Test as an indicator of your mastery of this topic.

13.1 The endocrine system regulates body functions with hormones

13.2 Hormones are classified as steroid or nonsteroid

13.3 The hypothalamus and the pituitary gland

Matching

____ 1. **endocrine system** a. an inactive molecule in a cell converted to an active molecule by the binding of a hormone to the cell membrane

____ 2. **hormones** b. cells in the hypothalamus that function as both nerve cells and endocrine cells

____ 3. **endocrine glands** c. a small region in the forebrain that plays an important role in homeostatic regulation

____ 4. **target cells** d. a communication system that uses chemical messengers

____ 5. **steroid hormones** e. a small endocrine gland called the master gland

____ 6. **nonsteroid hormones** f. groups of cells that secrete hormones into the extracellular space

____ 7. **second messenger** g. bloodborne units of information secreted by glands

____ 8. **hypothalamus** h. lipid-soluble hormones synthesized from cholesterol

____ 9. **pituitary gland** i. mostly lipid-insoluble hormones synthesized from amino acids

____ 10. **neuroendocrine cells** j. cells responsive to a particular hormone

Fill-in-the-Blank

Referenced sections are in parentheses.

11. Hormones circulate in the _____ and act only on specific _____. (13.1)

12. There are approximately _____ different hormones. (13.1)

13. The target cells of a hormone will have the right _____ for that particular hormone. (13.1)

14. Actions of the endocrine system are _____ than those of the nervous system. (13.1)

15. The _____ system and the _____ system interact with each other to maintain homeostasis. (13.1)

16. _____ hormones enter the cell to carry out their effects. (13.2)

17. _____ hormones may utilize a second messenger to carry out their effects. (13.2)

18. In a negative feedback loop involving a hormone, the effectors are the _____ _____. (13.2)

19. ADH and oxytocin are produced in the _____ and stored in the _____ _____. (13.3)

20. The release of hormones produced in the anterior pituitary is influenced by releasing or inhibiting _____ produced in the _____. (13.3)

21. The connection between the hypothalamus and the posterior pituitary is _____, while the connection between the hypothalamus and the anterior pituitary is _____. (13.3)

22. Diabetes insipidus may be caused by inadequate amounts of _____. (13.3)

23. In adults, an excess of growth hormone causes _____. (13.3)

Labeling

Using Figure 13.1 on page 174, write the name of each organ and the hormones secreted by that organ.

24. _____
25. _____
26. _____
27. _____
28. _____
29. _____
30. _____
31. _____

32. _____
33. _____
34. _____
35. _____
36. _____
37. _____
38. _____
39. _____

Figure 13.1

Word Choice

Circle the term or phrase that correctly completes each sentence.

40. Steroid hormones are synthesized from (cholesterol/amino acids).

41. Nonsteroid hormones are usually lipid (soluble/insoluble).

42. Nonsteroid hormones will (enter/not enter) a target cell.

43. Steroid hormones act in a cell by binding to (intracellular/cell surface) receptors.

44. Nonsteroid hormones may exert an effect on a target cell by influencing (gene expression/membrane ion channels).

45. Steroid hormones are (slower/faster) acting than nonsteroid hormones.

13.4 The pancreas secretes glucagon, insulin, and somatostatin

13.5 The adrenal gland comprises the cortex and medulla

13.6 Thyroid and parathyroid glands

13.7 The testes and ovaries produce sex hormones

13.8 Other glands and organs also secrete hormones

Matching

Match each hormone with the correct function.

____ 1. **glucagon** a. contributes to the action of the sympathetic nervous system

____ 2. **insulin** b. stimulates the production of red blood cells in the bone marrow

____ 3. **somatostatin** c. helps lymphocytes of the immune system mature into T cells

____ 4. **cortisol** d. lowers blood calcium levels by increasing the deposition of calcium in bone

____ 5. **aldosterone** e. helps to regulate blood pressure

____ 6. **epinephrine** f. raises blood sugar levels

____ 7. **thyroxine** g. regulates the development of sperm, the male reproductive organs, and male secondary sex characteristics

____ 8. **calcitonin** h. lowers blood sugar levels

____ 9. **parathyroid hormone** i. aids the functioning of the digestive system

____ 10. **testosterone** j. a glucocorticoid that helps maintain blood sugar levels by promoting the breakdown of fats and proteins

____ 11. **estrogen** k. stimulates aldosterone secretion and constricts blood vessels

_____ 12. **thymosin** l. a mineralocorticoid that regulates the amount of sodium and potassium in the body

_____ 13. **melatonin** m. inhibits the secretion of glucagon and insulin

_____ 14. **atrial natriuretic hormone** n. regulates the production of ATP in body cells

_____ 15. **cholecystokinin** o. initiates development of female secondary sex characteristics and helps regulate the menstrual cycle

_____ 16. **erythropoietin** p. increases blood calcium levels

_____ 17. **renin** q. may help synchronize body rhythms to the light/dark cycle

Crossword Puzzle

Write in the hormone that will be secreted in response to each physiological situation represented in each clue.

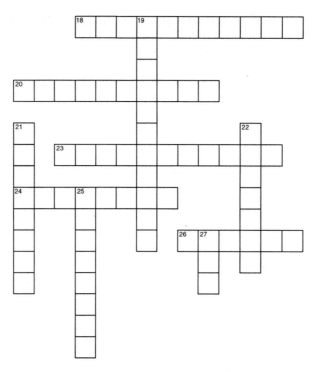

Across

18. _____ hormone is secreted when blood calcium levels are too low
20. blood calcium levels are too high
23. a sudden crisis activates the sympathetic nervous system
24. injury and emotional stress
26. _____ natriuretic hormone secreted when blood pressure rises

Down

19. potassium levels in the body are too high
21. blood sugar levels are too low
22. blood glucose levels increase
25. a situation that creates an increased need for energy
27. iodine deficiency leads to absence of thyroxin

13.9 Other chemical messengers

13.10 Disorders of the endocrine system

Completion

Complete the table below, writing in the chemical messenger associated with each action.

Chemical Messenger	Primary Action
1.	increases mucus secretion, dilates blood vessels, and increases the leakiness of capillaries
2.	locally control blood flow, contribute to inflammatory response, aid blood clotting
3.	regulates smooth muscle contraction in the digestive tract, controls penile erection
4.	influence cell division, influence direction of cell growth

For each of the endocrine disorders listed below, write in the cause and symptoms. Answers to this exercise are not listed in the answer key; refer to Section 13.10 in your text to check your answers.

5. Hypothyroidism:

6. Hyperthyroidism:

7. Addison's disease:

8. Cushing's syndrome:

continued

Chapter Test

Multiple Choice

1. Which of the following is *not* a defining feature of hormones?
   a. Hormones are secreted by endocrine glands.
   b. Hormones are secreted throughout life.
   c. Hormones circulate in the bloodstream.
   d. Hormones act only upon specific cells in the body.

2. Cells that respond to a certain hormone will have _____ specific for that hormone.
   a. receptors
   b. enzymes
   c. phospholipids
   d. glands

3. Endocrine control of body activities is usually _____ than nervous control.
   a. faster
   b. slower
   c. more specific
   d. less specific

4. The nervous system and the endocrine system interact through the:
   a. thyroid gland and the pituitary gland.
   b. pituitary gland and the hypothalamus.
   c. hypothalamus and the thyroid gland.
   d. anterior pituitary and posterior pituitary.

5. Steroid hormones exert their effect using:
   a. a second messenger.
   b. cyclic AMP.
   c. a hormone-receptor complex.
   d. a carbohydrate molecule.

6. Nonsteroid hormones:
   a. cross the cell membrane easily.
   b. bind to receptors on the cell surface.
   c. cross the membrane or bind to receptors, depending on the target cell.
   d. pass through protein channels in the membrane.

7. Nonsteroid hormones are usually faster acting than steroid hormones because:
    a. they are transported more quickly.
    b. they activate proteins that already exist within the cell.
    c. they utilize enzymes.
    d. they are not attacked by immune cells.

8. The hypothalamus is able to produce hormones because:
    a. it is in the brain.
    b. it contains neurosecretory cells.
    c. it is influenced by the pituitary gland.
    d. it is connected to the thalamus, which functions as a gland.

9. Oxytocin is produced by the _____ and stored in the _____.
    a. hypothalamus, posterior pituitary
    b. posterior pituitary, hypothalamus
    c. hypothalamus, anterior pituitary
    d. anterior pituitary, hypothalamus

10. Which of the following hormones is *not* produced by the anterior pituitary?
    a. adrenocorticotropic hormone
    b. luteinizing hormone
    c. prolactin
    d. antidiuretic hormone

11. Hormones produced in the anterior pituitary will be released in response to _____ produced by the _____.
    a. factors, thyroid
    b. releasing hormones, anterior pituitary
    c. releasing hormones, posterior pituitary
    d. releasing hormones, hypothalamus

12. Growth and development of the male and female reproductive organs is triggered by:
    a. FSH and LH.
    b. ACTH.
    c. prolactin.
    d. aldosterone.

13. Unusually high stature results from increased levels of _____ during childhood.
    a. testosterone
    b. growth hormone
    c. FSH and LH
    d. thyroxin

14. The _____ is both an endocrine and an exocrine gland.
    a. thyroid gland
    b. pituitary gland
    c. hypothalamus
    d. pancreas

15. Which of the following hormones will be most active after a 24-hour fast?
    a. insulin
    b. aldosterone
    c. glucagon
    d. somatostatin

16. Cortisol is produced by the:
    a. adrenal medulla.
    b. adrenal cortex.
    c. thyroid gland.
    d. thymus gland.

17. The adrenal medulla is similar to the hypothalamus because:
    a. they are both neuroendocrine organs.
    b. they are both divided into two regions.
    c. they both produce inhibiting and releasing hormones.
    d. they both respond to pituitary hormones.

18. The blood calcium regulatory function of which gland is most important in adulthood?
    a. thyroid gland
    b. parathyroid gland
    c. pancreas
    d. thymus gland

19. Goiter is caused by:
    a. increased levels of thyroxin.
    b. iodine deficiency.
    c. calcium deficiency.
    d. malfunction of the hypothalamus.

20. Which of the following is *not* a characteristic of parathyroid hormone?
    a. It increases blood calcium levels.
    b. It inhibits calcium retention by the kidneys.
    c. It increases calcium reabsorption by the digestive tract.
    d. It removes calcium from the bone.

21. Testosterone:
    a. is produced by the anterior pituitary.
    b. is produced at consistent levels throughout life.
    c. secretion is stimulated by LH.
    d. inhibits the production of aldosterone.

22. The kidneys secrete:
    a. secretin and cholecystokinin.
    b. renin and erythropoietin.
    c. histamine and prostaglandins.
    d. aldosterone and ADH.

23. Nitric oxide is not considered a hormone because:
    a. it is not a steroid.
    b. it is short lived.
    c. it is produced by epithelial tissue.
    d. it is not secreted into the bloodstream.

24. Which of the following sets of hormones oppose each other?
    a. secretin and renin
    b. FSH and LH
    c. glucagon and insulin
    d. progesterone and estrogen

25. The symptoms of Cushing's disease are caused by:
    a. lowered levels of cortisol and aldosterone.
    b. hypersecretion by the thyroid gland.
    c. excessive levels of cortisol.
    d. excessive levels of testosterone.

Key Concept Review Questions

Each of the Key Concepts listed at the beginning of this chapter has been rewritten as a question below. After successfully completing the study guide exercises and the Chapter Test, you should be able to answer each of these questions. Refer to the Key Concepts list at the beginning of this chapter to check your answers.

1. What structures are included in the endocrine system? What is the initial destination of secreted hormones?
2. On what cells do hormones act?
3. How does endocrine control of body activities compare to nervous control in terms of speed?
4. What are the two general classifications of hormones?
5. What are the characteristics of steroid hormones?
6. What are the characteristics of nonsteroid hormones?
7. What type of feedback loops do hormones participate in?
8. What gland allows interaction between the hypothalamus and the endocrine system?
9. What are the two regions of the pituitary gland?
10. What hormones are stored in the posterior pituitary? Where were these hormones produced?
11. What are the hormones produced by the anterior pituitary? What controls the release of these hormones?
12. What are the three hormones produced by the pancreas? What is the function of each hormone?
13. What hormones are produced by the adrenal cortex? What are the functions of these hormones?
14. What hormones are produced by the adrenal medulla? What stimulates release of these hormones?
15. What hormones are produced by the thyroid gland? What are the functions of these hormones?
16. What hormone is produced by the parathyroid gland? What is the function of this hormone?
17. What hormones are produced by the testes and the ovaries? What are the functions of these hormones?
18. What are the names and functions of the hormones produced by the thymus gland and pineal gland?
19. What are the names of four chemical messengers?
20. What are at least four disorders of the endocrine system discussed in this chapter?

The Digestive System and Nutrition

Chapter Summary and Key Concepts

After reading and studying this chapter you should know the following:

Sections 14.1, 14.2, 14.3, 14.4, 14.5

1. The digestive system provides nutrients to the body.
2. The walls of the gastrointestinal tract consist of four layers: the innermost mucosa, the submucosa, the muscularis, and the serosa.
3. Sphincters are thick rings of ciliary muscle that can close off the passageways between organs.
4. The five basic processes of the digestive system are mechanical processing and movement, secretion, digestion, absorption, and elimination.
5. Two types of motility important to the digestive system are peristalsis that moves food forward, and segmentation that mixes food.
6. In the mouth the teeth grind the food, and saliva begins the process of digestion.
7. The pharynx receives food from the mouth and initiates swallowing.
8. The esophagus is a muscular tube that connects the pharynx and the stomach.
9. The stomach functions to store food, digest food, and regulate the delivery of food to the small intestine.
10. The small intestine functions to digest carbohydrates, lipids, and proteins, and to absorb most of the nutrients.

Sections 14.6, 14.7, 14.8, 14.9

11. The accessory organs of the digestive system include the salivary glands, the pancreas, the liver, and the gallbladder.

12. The pancreas secretes digestive enzymes that complete digestion in the small intestine, and sodium bicarbonate that neutralizes stomach acid.

13. The liver produces bile, stores fat-soluble vitamins, stores glucose as glycogen, manufactures plasma proteins, synthesizes and stores some lipids, inactivates chemicals, converts ammonia to urea, and destroys worn-out red blood cells.

14. The gallbladder stores and releases bile.

15. The large intestine reabsorbs water and concentrates feces.

16. Proteins and carbohydrates are absorbed from the small intestine by active transport, lipids move by diffusion, and water moves by osmosis.

17. Digestion is regulated by the endocrine system and the nervous system.

Sections 14.10, 14.11, 14.12

18. Carbohydrates, lipids, and proteins are essential nutrients in the diet.

19. Body weight depends on the balance between energy consumed and energy used.

20. Disorders of the digestive system include lactose intolerance, diverticulosis, colon polyps, hepatitis, gallstones, malnutrition, anorexia nervosa, and bulimia.

Exercises

Complete the exercises for each section after you have read and studied the section. If you cannot answer some questions, or answer them incorrectly, return to the chapter and review this information. You may find it helpful to work on only one section at a time. When you have completed all sections, take the Chapter Test as an indicator of your mastery of this topic.

14.1 The digestive system brings nutrients into the body

14.2 The mouth processes food for swallowing

14.3 The pharynx and esophagus deliver food to the stomach

14.4 The stomach stores, digests, and regulates food delivery

14.5 The small intestine digests food and absorbs nutrients

Matching

___ 1. **digestive system** a. three glands near the oral cavity that produce saliva

___ 2. **gastrointestinal tract** b. secretory cells in the mucosa of the stomach

___ 3. **accessory organs** c. the region of the small intestine where most digestion occurs

___ 4. **peristalsis** d. a watery mixture of partially digested food and gastric juice that enters the small intestine

___ 5. **segmentation** e. a body system consisting of all the organs that function in getting nutrients into the body

___ 6. **mouth** f. capillaries and a lymph vessel at the center of a villus, which function to transport nutrients to larger vessels

___ 7. **salivary glands** g. a protein-digesting enzyme in the stomach

___ 8. **pharynx** h. muscular waves of contraction that propel food forward

___ 9. **esophagus** i. a mixture secreted by the stomach that contains HCl, pepsinogen, and fluid

___ 10. **stomach** j. a hollow tube formed by the hollow organs of the digestive system

___ 11. **gastric glands** k. an organ that receives chyme from the stomach, secretions from the pancreas, and bile from the gallbladder

___ 12. **pepsin** l. microscopic projections that cover large folds in the mucosa of the small intestine

___ 13. **gastric juice** m. the throat

___ 14. **chyme** n. a muscular tube that connects the pharynx to the stomach

___ 15. **small intestine** o. random contraction and relaxation of smooth muscle that mixes food

___ 16. **duodenum** p. a muscular sac that stores and digests food and regulates food delivery to the small intestine

___ 17. **villi** q. organs that aid the digestive system and are not part of the gastrointestinal tract

___ 18. **lacteal** r. location where digestion begins

Labeling

Label each indicated structure in Figure 14.1 with a name and a function.

Structures
a. stomach
b. pancreas
c. mouth
d. liver
e. pharynx

Functions
m. secretes saliva
n. food is chewed and moistened here
o. swallowing is initiated here
p. food mixes with gastric juice and protein digestion begins
q. produces bile

Structures
f. rectum
g. salivary glands
h. esophagus
i. small intestine
j. large intestine
k. anus
l. gallbladder

Functions
r. reabsorbs water and concentrates undigested food into feces
s. terminal opening for expelling feces
t. stores and concentrates bile
u. organ where digestion of all food is completed
v. secretes enzymes to digest all food groups
w. muscular tube that moves food from pharynx to stomach
x. passageway for fecal material

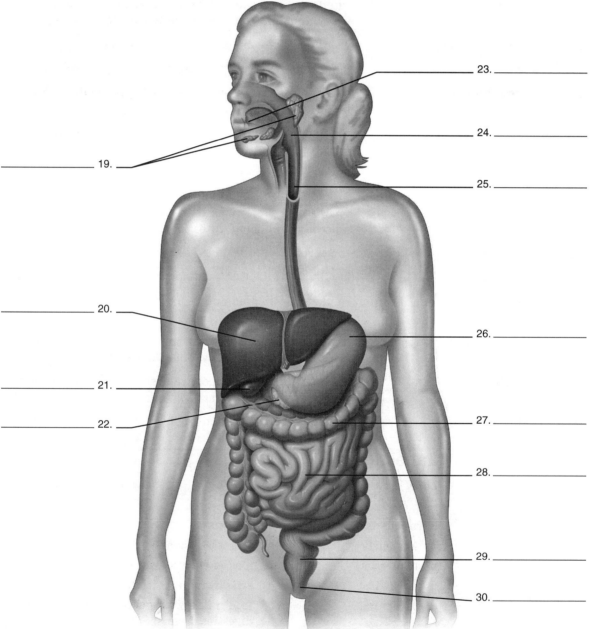

19. _____
20. _____
21. _____
22. _____
23. _____
24. _____
25. _____
26. _____
27. _____
28. _____
29. _____
30. _____

Figure 14.1

Fill-in-the-Blank

Referenced sections are in parentheses.

31. The space within the hollow tube of the GI tract is the _____. (14.1)

32. The muscularis consists of _____ muscle in a _____ and a _____ arrangement. (14.1)

33. The _____ is the entrance to the GI tract. (14.2)

34. The type of tooth specialized for cutting food is the _____. (14.2)

35. The visible region of a tooth is the _____. (14.2)

36. _____ is an enzyme found in saliva that begins the digestion of carbohydrates. (14.2)

37. Once initiated, the swallowing reflex is _____ and cannot be stopped. (14.3)

38. The _____ _____ is a circular muscle that lies between the stomach and the small intestine. (14.4)

39. Gastric glands release _____ _____ which consists of HCL, pepsinogen, and fluid. (14.4)

40. Stretching in the stomach signals the muscles to increase _____. (14.4)

41. The first region of the small intestine is the _____. (14.5)

42. The _____ and the _____ are segments of the small intestine where most absorption takes place. (14.5)

43. The surface of each epithelial cell of a villus is covered with smaller projections called _____. (14.4)

Labeling

Label each indicated structure in Figure 14.2 and then answer the questions below.

Figure 14.2

51. Which layer is responsible for motility of the digestive tract?
52. Which layer carries blood vessels and nerves?
53. Which layer protects and anchors the digestive tract?
54. Which layer contains smooth muscle?

14.6 Accessory organs aid digestion and absorption

14.7 The large intestine absorbs nutrients and eliminates waste

14.8 How nutrients are absorbed

Completion

For each function listed in the table, name the corresponding structure, organ, or substance. Words may be used more than once.

liver colon bile hepatic portal system

pancreas cecum bacteria sigmoid colon

Organ, Structure, or Substance	Function
1.	venous blood vessels that drain the stomach, intestines, and pancreas
2.	stores fat-soluble vitamins
3.	secretes proteases, lipases, and amylases into the duodenum
4.	is divided into ascending, transverse, and descending sections
5.	a pouch that receives chyme from the small intestine
6.	the portion of the colon that joins the rectum
7.	converts ammonia into urea
8.	a watery mixture of bicarbonate, cholesterol, and bile salts
9.	emulsifies lipids, breaking them down into smaller droplets
10.	produce vitamin K in the large intestine

Labeling

Use the terms listed below to label the structures and processes indicated in Figure 14.3.

bile salts   fat globules   diffusion

lacteal   chylomicrons   emulsion droplets

micelles   amino acids   free fatty acids and monoglycerides

exocytosis

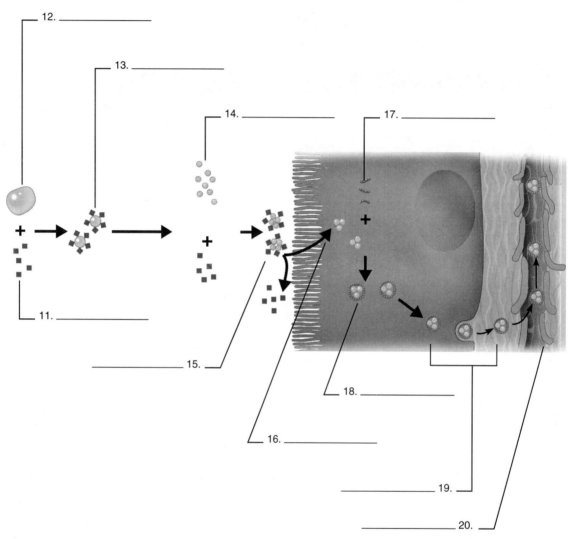

Figure 14.3

14.9 Endocrine and nervous systems regulate digestion

Fill-in-the-Blank

Using Figure 14.4, write in the correct term in each sentence below describing the regulation of digestion.

1. When the stomach stretches, neural reflexes increase _____ and secretion of _____ _____.

2. Stretching and the presence of protein stimulate the release of _____, which stimulates release of more _____ _____.

3. Acid in the chyme that enters the small intestine triggers release of _____, which stimulates the pancreas to secrete _____ to neutralize the acid.

4. The presence of fat and protein stimulate the release of _____, which signals the pancreas to secrete more _____ enzymes.

5. CCK and stretching of the duodenum stimulate the gallbladder to release _____.

6. Secretin, CCK, and stretching of the small intestine _____ stomach motility and secretions, preventing _____ from flowing too quickly.

7. Gastrin and a neural reflex involving stomach stretching increase _____ in the large intestine.

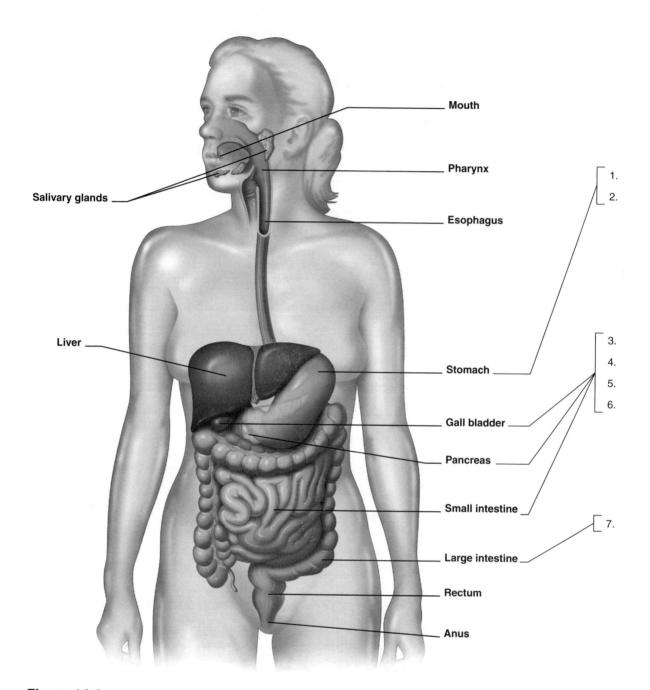

Figure 14.4

14.10   Nutrition: You are what you eat

14.11   Weight control: Energy consumed versus energy spent

14.12   Disorders of the digestive system

Matching

____ 1. **carbohydrates**   a. a protein that provides all 20 amino acids

____ 2. **complex carbohydrate**   b. noncancerous growths that project from the mucous membrane

____ 3. **lipids**   c. impairment of human development or function by overnutrition or undernutrition

____ 4. **saturated fat**   d. the energy the body uses to perform essential functions

____ 5. **proteins**   e. atoms of certain chemical elements present as ions that contribute to the function of bone, nerves, and muscles

____ 6. **essential amino acids**   f. a form of hepatitis passed in blood or body fluids

____ 7. **complete protein**   g. a fat that tends to be solid at room temperature

____ 8. **vitamins**   h. nutrient group that should comprise about 60% of daily calories

____ 9. **minerals**   i. binge-and-purge condition in which someone eats and then prevents calorie absorption

____ 10. **fiber**   j. inflammation of the liver

____ 11. **calorie**   k. excessive dieting to the point of starvation or death

____ 12. **basal metabolic rate**   l. nutrient group that should comprise about 10–30% of daily calories

____ 13. **diverticulosis**   m. hard crystals of cholesterol that obstruct bile flow

____ 14. **colon polyp**   n. the amount of energy needed to raise the temperature of 1 g of water 1°C

____ 15. **hepatitis**   o. an indigestible material that aids movement through the large intestine

____ 16. **hepatitis A**   p. a molecule that consists of many sugar units linked together, providing a long-lasting source of energy

____ 17. **hepatitis B**   q. nutrient group that should comprise about 15% of daily calories

____ 18. **gallstones**   r. a condition in which the mucosal lining of the large intestine protrudes through the intestinal wall

____ 19. **malnutrition**   s. a form of hepatitis transmitted by contaminated food or water

____ 20. **anorexia nervosa**   t. a group of at least 13 chemicals that are essential for normal functioning

____ 21. **bulimia**   u. amino acids that the body cannot synthesize and must be obtained from a food source

Chapter Test

Multiple Choice

1. The gastrointestinal tract is composed of:
   a. all the organs of the digestive system.
   b. the accessory organs of the digestive system.
   c. the hollow organs of the digestive system.
   d. the blood vessels and nerves that supply the digestive system.

2. Which layer of the gastrointestinal tract wall is connective tissue containing blood vessels, lymph vessels, and nerves?
   a. mucosa
   b. submucosa
   c. serosa
   d. muscularis

3. Food is propelled forward through the GI tract by:
   a. mechanical processing.
   b. peristalsis.
   c. segmentation.
   d. motility.

4. The innermost region of a tooth containing blood vessels and nerves is the:
   a. enamel.
   b. crown.
   c. pulp cavity.
   d. dentin.

5. The tongue is primarily:
   a. skeletal muscle.
   b. smooth muscle.
   c. connective tissue.
   d. epithelial tissue.

6. Which of the following is *not* a component of saliva?
   a. mucin
   b. bicarbonate
   c. lysozyme
   d. proteinase

7. Which of the following represents the path taken by food after leaving the mouth?
   a. pharynx, larynx, esophagus, small intestine, large intestine
   b. pharynx, esophagus, stomach, small intestine, large intestine
   c. larynx, esophagus, stomach, large intestine, small intestine
   d. larynx, esophagus, stomach, small intestine, large intestine

8. Malfunction of the lower esophageal sphincter may result in:
   a. bulimia.
   b. acid reflux.
   c. stomach ulcer.
   d. diarrhea.

9. Gastric glands secrete:
   a. HCl and pepsinogen.
   b. gastrin and secretin.
   c. gastrin and lipases.
   d. intrinsic factor and amylase.

10. Enzyme activity in the stomach contributes to the digestion of:
    a. carbohydrates.
    b. lipids.
    c. proteins.
    d. all of the above

11. Most of the digestion that takes place in the small intestine occurs in the:
    a. duodenum.
    b. ileum.
    c. jejunum.
    d. lacteals.

12. The surface area of the small intestine is greatly increased by the presence of:
    a. serosa and submucosa.
    b. muscularis and epithelial cells.
    c. duodenum, jejunum, and ileum.
    d. villi and microvilli.

13. The pancreas secretes enzymes that aid the digestion of:
    a. carbohydrates.
    b. lipids.
    c. proteins.
    d. all of the above

14. Bile is produced by the _____ and stored in the _____.
    a. gallbladder, small intestine
    b. gallbladder, liver
    c. liver, small intestine
    d. liver, gallbladder

15. Nutrient-rich blood leaving the GI tract goes directly to the:
    a. large intestine.
    b. vessels of the body.
    c. liver.
    d. kidneys.

16. The large intestine absorbs primarily:
    a. nutrients.
    b. vitamins and minerals.
    c. lipids.
    d. water.

17. Which of the following represents the correct sequence in which structures involved in lipid absorption are formed?
    a. micelles, triglycerides, chylomicrons
    b. triglycerides, chylomicrons, micelles
    c. micelles, chylomicrons, triglycerides
    d. chylomicrons, micelles, triglycerides

18. Proteins and carbohydrates are absorbed from the small intestine by:
    a. diffusion.
    b. facilitated diffusion.
    c. active transport.
    d. osmosis.

19. Most digestive regulatory processes involve the:
    a. mouth and stomach.
    b. stomach and large intestine.
    c. stomach and small intestine.
    d. small intestine and large intestine.

20. Complex carbohydrates are better nutritionally than simple carbohydrates because:
    a. simple carbohydrates contain more fats.
    b. simple carbohydrates contain more calories.
    c. complex carbohydrates provide a more lasting source of energy.
    d. complex carbohydrates are easier to absorb.

21. Essential amino acids are those that:
    a. are required for protein synthesis.
    b. the body is capable of synthesizing.
    c. the body is not capable of synthesizing.
    d. are quickly broken down in the body.

22. Water-soluble vitamins:
    a. need to be consumed on a regular basis.
    b. are rapidly excreted in the urine.
    c. are stored in the liver.
    d. a and b
    e. all of the above

23. Your basal metabolic rate is:
    a. the number of calories you consume each day.
    b. the number of calories you use each day.
    c. the number of calories your body needs to perform essential activities.
    d. the number of calories your body needs to perform nonessential activities.

24. To lose 1 pound of fat, you must spend:
    a. 1,000 calories.
    b. 1,500 calories.
    c. 2,500 calories.
    d. 3,500 calories.

25. A form of hepatitis transmitted by contaminated food or water is:
    a. hepatitis A.
    b. hepatitis B.
    c. hepatitis C.
    d. hepatitis E.

Key Concept Review Questions

Each of the Key Concepts listed at the beginning of this chapter has been rewritten as a question below. After successfully completing the study guide exercises and the Chapter Test, you should be able to answer each of these questions. Refer to the Key Concepts list at the beginning of this chapter to check your answers.

1. What is the function of the digestive system?
2. What are the four layers in the wall of the gastrointestinal tract?
3. What are sphincters composed of, and what is their function?
4. What are the five basic processes of the digestive system?
5. What are two types of motility that function in the digestive system?
6. What occurs in the mouth?
7. What happens in the pharynx?
8. What two organs are connected by the esophagus?

9. What are the functions of the stomach?
10. What are the functions of the small intestine?
11. What are the four accessory organs of the digestive system?
12. What are the functions of the pancreas?
13. What are the functions of the liver?
14. What are the functions of the gallbladder?
15. What are the functions of the large intestine?
16. What transport processes are used in the absorption of carbohydrates, proteins, lipids, and water?
17. What two body systems are involved in the regulation of the digestive system?
18. What are the names of three classes of compounds that are essential nutrients in the diet?
19. How does energy consumption and energy usage affect body weight?
20. What are eight disorders of the digestive system discussed in this chapter?

15

The Urinary System

Chapter Summary and Key Concepts

After reading and studying this chapter you should know the following:

Sections 15.1, 15.2, 15.3

1. Excretion refers to processes that remove wastes and excess materials from the body.

2. Excretion is accomplished by the digestive system, the lungs, the skin, the liver, and the urinary system.

3. The urinary system consists of the kidneys, ureters, bladder, and urethra.

4. The urinary system maintains a constant internal environment by regulating water balance and body levels of nitrogenous wastes, ions, and other substances.

5. The main organs of the urinary system are the kidneys.

6. The kidneys form urine and consist of the outer cortex, the inner medulla, and the renal pelvis.

7. The ureters transport urine to the bladder, the bladder stores urine, and the urethra carries urine from the body.

8. The nephron is the functional unit of the kidney.

9. Blood supply to the nephron is provided by the afferent arterioles, peritubular capillaries, and vasa recta.

Sections 15.4, 15.5, 15.6

10. The formation of urine involves three processes: glomerular filtration, tubular reabsorption, and tubular secretion.

11. The glomerular filtrate is produced by high blood pressure in the glomerular capillaries, and it contains water and small solutes in the same concentration as blood plasma.

12. Tubular reabsorption returns filtered water, sodium, major nutrients, and bicarbonate to the blood.

13. Tubular secretion moves substances from the capillaries into the tubules.

14. The kidneys can excrete excess water by producing dilute urine and can conserve water by producing concentrated urine.

Sections 15.7, 15.8

15. The kidneys contribute to homeostasis by adjusting blood volume and pressure, regulating salt balance, maintaining blood pH, controlling production of red blood cells, and participating in the formation of vitamin D.

16. Antidiuretic hormone influences water balance by increasing the water permeability of the collecting duct.

17. Blood volume is influenced by aldosterone, renin, and atrial natriuretic hormone.

18. The kidneys maintain blood pH by secreting H^+.

19. The kidneys stimulate red blood cell production by secreting erythropoietin.

20. Disorders of the urinary system include kidney stones, acute and chronic renal failure, and urinary tract infections.

Exercises

Complete the exercises for each section after you have read and studied the section. If you cannot answer some questions, or answer them incorrectly, return to the chapter and review this information. You may find it helpful to work on only one section at a time. When you have completed all sections, take the Chapter Test as an indicator of your mastery of this topic.

15.1 The urinary system contributes to homeostasis

15.2 Organs of the urinary system

15.3 Nephrons produce urine

Matching

____ 1. **excretion** a. the portion of a nephron that surrounds and encloses the glomerulus

____ 2. **urinary system** b. a muscular tube that transports urine to the bladder

____ 3. **urea** c. a vessel that carries blood away from the glomerulus

____ 4. **kidneys** d. blood vessels that supply the loop of Henle and collecting duct

____ 5. **ureter** e. portion of the nephron that is affected by aldosterone

_____ 6. **urinary bladder**   f. a vessel that carries blood into the nephron

_____ 7. **urethra**   g. processes that remove wastes and excess materials from the body

_____ 8. **nephron**   h. a tube that receives filtrate from many nephrons

_____ 9. **Bowman's capsule**   i. a capillary network formed by the efferent arterioles that carries away reabsorbed water, ions, and nutrients

_____ 10. **proximal tubule**   j. a network of capillaries that produces the filtrate

_____ 11. **loop of Henle**   k. the primary organs of the urinary system

_____ 12. **distal tubule**   l. the site of the most tubular reabsorption

_____ 13. **collecting duct**   m. the major nitrogenous waste product in urine

_____ 14. **afferent arteriole**   n. a muscular tube that carries urine from the body

_____ 15. **glomerulus**   o. the portion of the nephron that extends deep into the medulla

_____ 16. **efferent arterioles**   p. a body system that produces, transports, stores, and excretes urine

_____ 17. **peritubular capillaries**   q. the functional unit of the kidney

_____ 18. **vasa recta**   r. an organ that stores urine

Labeling

Use the terms below to label Figure 15.1.

kidney   ureter   renal pelvis
medulla   cortex   renal blood vessels
ureter   bladder   major blood vessels
nephron   urethra

Figure 15.1

Labeling

Use the terms below to label Figure 15.2.

glomerulus
artery
glomerular capsule
ascending limb
efferent arteriole
loop of Henle
distal tubule
vein
proximal tubule
peritubular capillaries
afferent arteriole
collecting duct
descending limb

Figure 15.2

15.4 Formation of urine—filtration, reabsorption, and secretion

15.5 The kidneys can produce dilute or concentrated urine

15.6 Urination depends on a reflex

Fill-in-the-Blank

Referenced sections are in parentheses.

1. The three processes of urine formation are _____ _____, _____ _____, and _____ _____. (15.4)

2. The glomerular filtrate is filtered from the _____ _____ and flows into the _____ _____. (15.4)

3. Glomerular filtrate contains water and small _____, but not large _____ or blood _____. (15.4)

4. The driving force for glomerular filtration is high _____ _____ in the glomerular capillaries. (15.4)

5. During tubular reabsorption, water and solutes move from the _____ into the _____ _____. (15/4)

6. Most tubular reabsorption occurs in the _____ tubule. (15.4)

7. Substances that are almost completely reabsorbed include _____, _____, _____, and _____. (15.4)

8. The distal tubule and collecting duct reabsorb less than _____ of the water, but this is the region where water excretion is _____. (15.4)

9. During tubular reabsorption, the _____ _____ of Na$^+$ is the driving force for the reabsorption of equal proportions of sodium, chloride, water, and nutrients.

10. The process of tubular secretion moves substances from the _____ into the _____ to be _____. (15.4)

11. The proximal tubule secretes _____ and _____. (15.4)

12. The distal tubule secretes _____.

13. A dilute urine is produced when the kidneys reabsorb _____ water. (15.5)

14. The ability to form dilute urine depends partly on the _____ of solutes in the _____ _____. (15.5)

15. The solute concentration of the renal medulla is kept high by the recycling of _____. (15.5)

16. A concentrated urine is produced when the kidneys _____ more water. (15.5)

17. The reabsorption of water is regulated by _____. (15.5)

18. ADH _____ the permeability of the _____ _____ to water. (15.5)

19. In the presence of ADH, more _____ is reabsorbed, producing a _____ urine. (15.5)

20. Urination depends on a neural reflex called the _____ _____. (15.6)

21. The internal urethral sphincter consists of _____ muscle, and the external urethral sphincter consists of _____ muscles. (15.6)

Completion

Name the region of the nephron where each of the following events occurs.

Region	Event
22.	Water does not move out of or into the tubule because this region is impermeable to water.
23.	NaCl is actively transported out of the tubule.
24.	
25.	Urea diffuses into the tubule.
26.	NaCl diffuses out of the tubule.
27.	ADH increases the water permeability of this region.
28.	A countercurrent mechanism that involves these two regions.
29.	

15.7 The kidneys maintain homeostasis in many ways
15.8 Disorders of the urinary system

Labeling

1. Fill in Figure 15.3, outlining the regulation of blood volume.

   angiotensinogen   vasoconstriction   ADH
   aldosterone   angiotensin I   water
   renin   blood pressure   salt

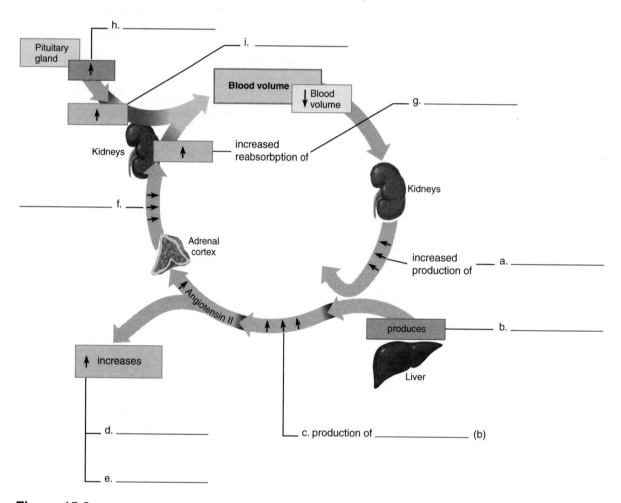

Figure 15.3

Short Answer

Answer the following questions about the role of the kidneys in maintaining homeostasis.

2. What change in blood solute concentration acts as a stimulus for increased secretion of ADH from the posterior pituitary?

3. What is the effect of aldosterone on the distal tubule and collecting duct?

4. What is the strongest stimulus for aldosterone secretion?

5. What cells secrete renin?

6. What organ secretes atrial natriuretic hormone?

7. What is the action of atrial natriuretic hormone (ANH)?

8. ANH has an effect opposite to that of what two hormones?

Paragraph Completion

Use the following terms to complete the paragraphs, then repeat this exercise with the terms covered. Terms used more than once are listed multiple times.

H^+	decrease	vitamin D	excreted
kidneys	liver	ammonia	H^+
increase	bicarbonate	calcium	decreases
phosphorus	UV rays	erythropoietin	

The kidneys help to maintain homeostasis in the body in many ways. In addition to monitoring blood volume and blood pressure, they influence acid-base balance, red blood cell production, and vitamin D synthesis. The kidneys help to maintain acid-base balance by secreting (9) _____ into the tubule, and by producing (10) _____. The hydrogen ions aid the reabsorption of (11) _____, and ammonia combines with excess (12) _____ and is (13) _____.

(14) _____ is a hormone produced by the kidneys when oxygen levels in the kidney (15) _____. Erythropoietin stimulates the bone marrow to (16) _____ red blood cell production. As oxygen levels return to normal, erythropoietin production (17) _____.

continued

The kidneys also participate in (18) _____ synthesis. Vitamin D is necessary for the absorption of (19) _____ and (20) _____ from the digestive tract. Vitamin D synthesis begins in the skin when exposure to (21) _____ initiates the production of an inactive form of vitamin D. This inactive vitamin D precursor is transported first to the (22) _____ and then to the (23) _____ where it is converted to a useable form.

Completion

Name each disorder of the urinary system or treatment described in the table.

Disorder or Treatment	Description
24.	crystallization of minerals in the urine, forming solid masses that can block urine flow
25.	kidney impairments that are short-term and possibly correctable
26.	long-term irreversible damage to the kidneys
27.	treatment in which a patient's blood is passed through a semipermeable membrane
28.	treatment in which dialysis fluid is placed directly into the peritoneal cavity

Chapter Test

Multiple Choice

1. Which of the following is not part of the urinary system?
   a. kidneys
   b. liver
   c. bladder
   d. ureters

2. The major nitrogenous waste product in urine is:
   a. urea.
   b. creatinine.
   c. bicarbonate.
   d. salt.

3. The hollow space of a kidney where urine collects after it is formed is the:
   a. renal medulla.
   b. renal cortex.
   c. renal collecting duct.
   d. renal pelvis.

4. A muscular tube that extends from the bladder to the body surface is the:
   a. urethra.
   b. ureter.
   c. collecting duct.
   d. excretion tube.

5. The functional unit of the kidney is the:
   a. renal medulla.
   b. pyramid.
   c. nephron.
   d. glomerulus.

6. The structure of the glomerular capillaries contributes to the formation of filtrate because:
   a. the walls of the capillaries are strengthened by smooth muscle to withstand the high blood pressure in the glomerulus.
   b. the walls of the capillaries contain tight junctions, preventing excess plasma from leaking into the glomerulus.
   c. the walls of the capillaries are much more permeable to water and small solutes than other capillaries.
   d. the branching of the capillaries provides for even distribution of blood pressure.

7. The ascending and descending limbs are parts of the:
   a. proximal tubule.
   b. glomerulus.
   c. efferent arteriole.
   d. loop of Henle.

8. The _____ arteriole carries blood into the glomerulus, and the _____ arteriole carries blood away.
   a. proximal, distal
   b. ascending, descending
   c. filtering, collecting
   d. afferent, efferent

9. Substances that are reabsorbed from the proximal tubule will enter the:
   a. afferent arteriole.
   b. peritubular capillaries.
   c. collecting duct.
   d. descending arteriole.

10. Most tubular reabsorption occurs in the:
    a. proximal tubule.
    b. loop of Henle.
    c. collecting duct.
    d. renal medulla.

11. Which of the following is not almost completely reabsorbed from the filtrate?
    a. glucose
    b. water
    c. creatinine
    d. salt

12. Sodium is reabsorbed from the proximal tubule by:
    a. passive diffusion.
    b. facilitated diffusion.
    c. active transport.
    d. osmosis.

13. What two ions are secreted by the proximal tubule in maintaining acid-base balance?
    a. hydrogen and bicarbonate
    b. urea and ammonium
    c. hydrogen and ammonium
    d. potassium and hydrogen

14. The diffusion of water out of the descending limb creates a gradient that favors:
    a. the diffusion of water into the ascending limb.
    b. the diffusion of NaCl out of the ascending limb.
    c. the active transport of glucose out of the ascending limb.
    d. the active transport of H^+ into the descending limb.

15. The action of ADH on the collecting duct produces:
    a. a dilute urine.
    b. a concentrated urine.
    c. urine with a high glucose content.
    d. the presence of protein in the urine.

16. Which urethral sphincter is voluntary?
    a. internal
    b. external
    c. secondary
    d. primary

17. The juxtaglomerular apparatus consists of cells of the:
    a. afferent arteriole and distal tubule.
    b. afferent arteriole and proximal tubule.
    c. efferent arteriole and distal tubule.
    d. efferent arteriole and collecting duct.

18. Caffeine acts as a mild diuretic because it:
    a. stimulates the release of ADH.
    b. stimulates sodium reabsorption.
    c. inhibits sodium reabsorption.
    d. increases the permeability of the glomerular capillaries.

19. The effect of aldosterone on the distal tubule and collecting duct is to:
    a. increase reabsorption of sodium.
    b. decrease reabsorption of sodium.
    c. increase excretion of water.
    d. increase secretion of H^+.

20. The greatest stimulus for release of aldosterone is:
    a. high sodium concentrations in the filtrate.
    b. decreased blood volume.
    c. increased blood volume.
    d. increased release of renin.

21. Renin is produced by:
    a. afferent arteriole cells in the juxtaglomerular apparatus.
    b. the liver.
    c. the pituitary.
    d. efferent arteriole cells in the juxtaglomerular apparatus.

22. The kidneys regulate the production of red blood cells by secreting:
    a. renin.
    b. erythropoietin.
    c. angiotensin I.
    d. vitamin D.

23. If the liver were unable to produce angiotensinogen, which of the following might occur?
    a. the kidneys would be unable to produce renin
    b. ADH levels would increase
    c. blood vessels would dilate
    d. angiotensin II would not be available

24. What is the role of the kidneys in the formation of vitamin D?
    a. the kidneys function alone in the synthesis of vitamin D
    b. the kidneys function with the skin and the liver in vitamin D synthesis
    c. the kidneys increase vitamin D secretion when blood pressure is low
    d. the kidneys secrete vitamin D and erythropoietin together

25. What is "end-stage renal disease"?
    a. chronic kidney failure
    b. acute kidney failure
    c. persistent kidney stones
    d. a condition that occurs when hemodialysis is no longer effective

Key Concept Review Questions

Each of the Key Concepts listed at the beginning of this chapter has been rewritten as a question below. After successfully completing the study guide exercises and the Chapter Test, you should be able to answer each of these questions. Refer to the Key Concepts list at the beginning of this chapter to check your answers.

1. What is excretion?

2. What systems and organs contribute to excretion?

3. What are the organs of the urinary system? What are the functions of the urinary system?

4. How does the urinary system contribute to homeostasis?

5. What are the main organs of the urinary system?

6. What are the three main regions of the kidney?

7. How do the ureters, the bladder, and the urethra contribute to the function of the urinary system?

8. What is the functional unit of the kidney?

9. What vessels supply blood to a nephron?

10. What are the three processes involved in the formation of urine?

11. What pressure contributes to the formation of glomerular filtrate, and what does the filtrate contain?

12. What occurs in tubular reabsorption?

13. What occurs in tubular secretion?

14. What changes in water excretion do the kidneys make to produce dilute or concentrated urine?

15. What are five ways in which the kidneys contribute to homeostasis?

16. How does ADH influence water balance?

17. What are three hormones or chemicals that influence blood volume?

18. How do the kidneys maintain blood pH?

19. How do the kidneys stimulate red blood cell production?

20. What four disorders of the urinary system are discussed in this chapter?

16

Reproductive Systems

Chapter Summary and Key Concepts

After reading and studying this chapter you should know the following:

Section 16.1

1. The reproductive system consists of primary reproductive organs, accessory glands, and ducts.

2. The primary organs of the male reproductive system are the testes, which secrete testosterone and produce sperm.

3. Sperm are produced in the seminiferous tubules and mature in the epididymis.

4. Sperm leave the body through the vas deferens and urethra in a fluid called semen.

5. Semen contains sperm and glandular secretions that provide nutrients, alkaline fluid, and mucus.

6. Sperm cells contain 23 chromosomes and are produced by a cell division process called meiosis.

7. Testosterone, LH, and FSH regulate the male reproductive system.

Section 16.2

8. The primary organs of the female reproductive system are the ovaries, which release oocytes.

9. The oviduct is a hollow tube that transports the oocyte from the ovary to the uterus, and it is the site of fertilization.

10. The uterus supports the development of a fertilized egg, and it consists of an inner endometrium and an outer myometrium.

11. The vagina is the female organ of intercourse and the birth canal.

12. The mammary glands are accessory organs that produce and store milk.

Chapter 16 Reproductive Systems

Sections 16.3, 16.4, 16.5

13. The ovaries and uterus undergo a monthly pattern of changes called the menstrual cycle, which consists of an ovarian cycle and a uterine cycle.

14. The ovarian cycle results in the release of a secondary oocyte capable of being fertilized.

15. The uterine cycle prepares the endometrium of the uterus to accept and nurture a fertilized egg.

16. GnRH, LH, FSH, estrogen, and progesterone are hormones involved in regulating the female reproductive system.

17. The human sexual response is a series of events that coordinate sexual function to accomplish intercourse and fertilization.

18. Birth control methods are designed to prevent fertilization.

19. Disorders of the reproductive system include infertility, complications of pregnancy, cancers and tumors of the reproductive organs, and sexually transmitted diseases.

20. STDs caused by bacteria are curable, while those caused by viruses are not. STDs include gonorrhea, syphilis, chlamydia, AIDS, hepatitis B, herpes, and genital warts.

Exercises

Complete the exercises for each section after you have read and studied the section. If you cannot answer some questions, or answer them incorrectly, return to the chapter and review this information. You may find it helpful to work on only one section at a time. When you have completed all sections, take the Chapter Test as an indicator of your mastery of this topic.

16.1 Male reproductive system delivers sperm

Matching

_____ 1. **reproductive system**   a. male organ of sexual intercourse

_____ 2. **sperm**   b. cells in the seminiferous tubules that nourish and aid the development of sperm cells

_____ 3. **egg**   c. site of sperm maturation and storage

_____ 4. **testes**   d. cells in the seminiferous tubules that secrete testosterone

_____ 5. **scrotum**   e. a gland that produces an alkaline fluid contained in semen

_____ 6. **seminiferous tubules**   f. tissues and organs that participate in creation of a new organism

_____ 7. **vas deferens**   g. fluid containing sperm and secretions from the seminal vesicle, prostate gland, and bulbourethral gland

_____ 8. **Sertoli cells**   h. male gamete

_____ 9. **epididymis**   i. gland that secretes fructose and most of the seminal fluid

_____ 10. **penis**   j. a gland that secretes lubricating mucus

_____ 11. **erection**   k. site of sperm formation in the testes

_____ 12. **semen**   l. skin and smooth muscle that enclose the testis

_____ 13. **seminal vesicle** m. male sex hormone

_____ 14. **prostate gland**   n. female gamete

_____ 15. **bulbourethral gland** o. an increase in length, diameter, and stiffness of the penis

_____ 16. **testosterone**   p. a tube that leads from the epididymis to the urethra

_____ 17. **Leydig cells**   q. a male reproductive organ that produces sperm, testosterone, and inhibin

Fill-in-the-Blank

18. The seminal vesicles produce seminal fluid containing _____ and _____.

19. Undifferentiated cells that will give rise to sperm are called _____.

20. In humans, diploid cells have _____ chromosomes, and haploid cells have _____ chromosomes.

21. Secretion of FSH by the anterior pituitary is inhibited by the hormone _____.

22. Primary spermatocytes are produced by a cell division process called _____.

23. The meiotic division of secondary spermatocytes produces _____.

24. The mitochondria in a sperm produce energy and are located in the _____.

25. Blood levels of testosterone are maintained by a _____ feedback loop involving the hypothalamus and the _____ _____.

26. Decreased levels of testosterone stimulate the hypothalamus to secrete _____.

27. _____ released by the pituitary increases secretion of testosterone.

Chapter 16 *Reproductive Systems* **217**

Labeling

Use the terms below to label Figure 16.1.

urethra	seminal vesicle	prostate gland
testis	vas deferens	epididymis
penis	seminiferous tubules	bulbourethral gland

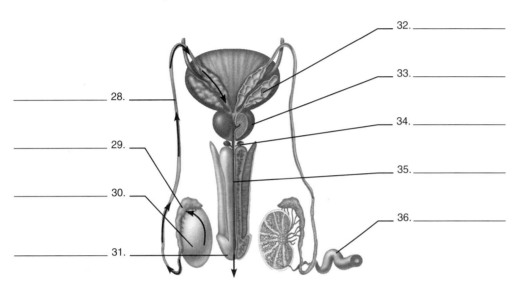

Figure 16.1

16.2 Female reproductive system produces eggs and supports pregnancy

Matching

_____ 1. **ovaries** a. the female external genitalia

_____ 2. **oocytes** b. the production of milk by the mammary glands

_____ 3. **estrogen and progesterone** c. a small organ that contains erectile tissue

_____ 4. **oviduct** d. the birth canal

_____ 5. **uterus** e. female reproductive organs, site of storage and development of oocytes

_____ 6. **endometrium** f. modified sweat glands containing milk-producing lobules

_____ 7. **cervix** g. female sex hormones

_____ 8. **vagina** h. the outer skin folds of the vulva

_____ 9. **vulva** i. a hollow organ that supports the growth and development of a fertilized egg

_____ 10. **labia majora** j. female gametes

_____ 11. **labia minora** k. the inner layer of the uterus

_____ 12. **clitoris**   l. a tube leading from the ovary to the uterus
_____ 13. **mammary glands** m. narrow end of the uterus, which leads into the vagina
_____ 14. **lactation**   n. the smaller inner folds of the vulva

Fill-in-the-Blank

15. Fertilization occurs in the _____.

16. In the process of implantation, a fertilized egg attaches to the _____.

17. The _____ is the pigmented area of the breast that surrounds the nipple.

18. _____ glands in the breast are specialized to produce milk.

19. The hormone _____ stimulates lactation, and the hormone _____ stimulates the release of milk.

Labeling

Use the terms below to label Figure 16.2.

clitoris	uterus	vagina	labium minora
oviduct	ovary	fimbriae	bladder
labium majora	urethra	cervix	

20. _____
21. _____
22. _____
23. _____
24. _____
25. _____
26. _____
27. _____
28. _____
29. _____
30. _____

Figure 16.2

16.3 The menstrual cycle consists of ovarian and uterine cycles

16.4 Human sexual response, intercourse, and fertilization

16.5 Birth control methods: Controlling infertility

Completion

Use the terms below to complete the paragraph, then repeat this exercise with the terms covered.

endometrium	secondary	pellucida	ovulation
ovarian	Meiosis II	corpus	Graafian
antrum	anterior	LH	puberty
FSH	ovarian	follicle	decrease
granulosa	primary	Meiosis I	zona
pituitary	ovulation	GnRH	hypothalamus
progesterone	luteum	glycoprotein	Meiosis I

The (1) _____ cycle involves the maturation and release of oocytes from the ovary. Each ovary contains approximately one million (2) _____ oocytes at birth. The oocytes have already begun the process of (3) _____. Continued development of an oocyte does not begin until (4) _____, and an oocyte will not complete (5) _____ until a sperm enters.

Each primary oocyte is surrounded by nourishing cells called (6) _____ cells. These cells, along with the oocyte, make up an immature (7) _____. About a dozen immature follicles begin the development process each month in response to (8) _____ and LH released by the (9) _____ (10) _____. At the beginning of the ovarian cycle, (11) _____ from the (12) _____ increases the concentration of these hormones, triggering oocyte development. Usually only one oocyte will complete the maturation process and be released during (13) _____.

As an immature follicle grows, the granulosa cells begin to secrete (14) _____ that become a noncellular coating called the (15) _____ (16) _____. Next, a fluid-filled space called the (17) _____ develops, and the follicle begins secreting estrogen and (18) _____. The primary ooctye will now finish the process of (19) _____, producing a (20) _____ oocyte and a polar body. This mature follicle is now called a (21) _____ follicle.

As estrogen (produced by the follicle) levels rise, the pituitary is stimulated to release a surge of the hormone (22) _____. This hormone surge triggers the process of (23) _____, the follicle ruptures, and the secondary oocyte is released. The remnants of the follicle will now form the (24) _____ (25) _____, which secretes large amounts of estrogen and progesterone to prepare the (26) _____ for implantation, if fertilization should occur. If fertilization does not occur, the corpus luteum degenerates, estrogen and progesterone levels (27) _____, and the (28) _____ cycle will begin again.

Short Answer

Refer to Figure 16.3 of the menstrual cycle to answer the questions.

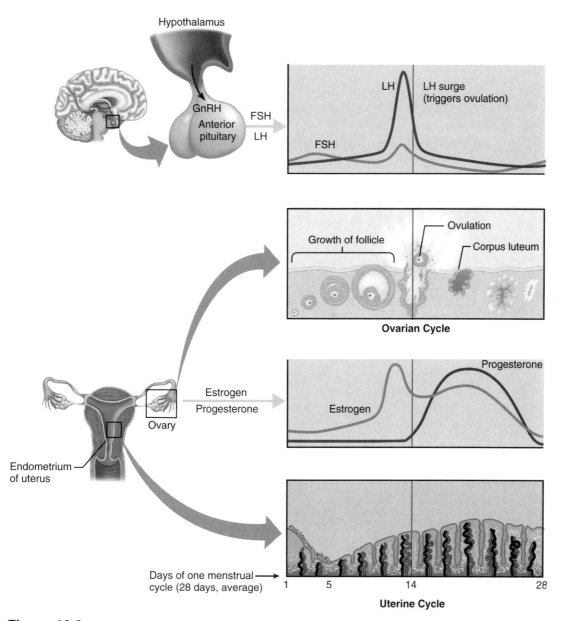

Figure 16.3

29. What days constitute the menstrual phase?

30. What body levels of estrogen and progesterone contribute to the events of the menstrual phase?

31. What days constitute the proliferative phase?

continued

32. What hormone activity triggers the proliferative phase?

33. What event of the ovarian cycle coincides with the end of the proliferative phase?

34. What uterine phase covers the last half of the menstrual cycle?

35. What hormone level is the highest during the last half of the menstrual cycle?

36. What event is the uterus preparing for during this phase?

37. In the absence of fertilization, what event in the ovarian cycle causes the levels of estrogen and progesterone to decline?

38. Does positive feedback influence the menstrual cycle before or after ovulation?

39. Describe this positive feedback loop.

40. What prevents development of a second follicle during the luteal phase?

Completion

Name each birth control method described in the table.

Method of Birth Control	Description
41.	the vas deferens is tied in two places and the segment between the ties is removed
42.	consists of an injection of progesterone that lasts 3 months, inhibiting FSH and LH
43.	a chemical in the form of a foam, cream, jelly, or douche that destroys sperm

Method of Birth Control	Description
44.	a plastic or metal piece inserted into the uterus that causes inflammation, interfering with implantation
45.	intercourse is avoided from about 5 days before ovulation until 3 days after ovulation
46.	a monthly injection of estrogen and progesterone
47.	a patch that slowly releases hormones into the bloodstream
48.	blocks the action of progesterone, causing a chemical abortion

16.6 Infertility: Inability to conceive

16.7 Sexually transmitted diseases: A worldwide concern

Matching

____ 1. **endometriosis** a. a bacterial infection and currently the most common STD

____ 2. **pelvic inflammatory disease** b. this STD is caused by a protozoan, and causes a discharge in both males and females

____ 3. **infertility** c. sperm are placed into the vagina or uterus close to the time of ovulation

____ 4. **in vitro fertilization** d. symptoms of this viral infection include genital blisters and painful urination; no cure is currently available

____ 5. **gamete intrafallopian transfer** e. extensive bacterial infection of the female reproductive tract

____ 6. **gonorrhea** f. failure to achieve pregnancy after trying for 1 year

____ 7. **chlamydia** g. STD caused by the human papilloma virus; no cure is currently available

____ 8. **genital herpes** h. unfertilized eggs and sperm are placed in the oviduct

____ 9. **genital warts** i. 50% of women contract this STD after a single exposure, and up to 80% of women may have no symptoms

____ 10. **trichomoniasis** j. fertilization occurs in a test tube, and the embryo is inserted into the uterus

____ 11. **artificial insemination** k. growth of endometrial tissue on organs outside the ovaries

Chapter Test

Multiple Choice

1. Which of the following is a secondary sexual characteristic?
   a. testosterone
   b. testis
   c. seminal vesicle
   d. muscle mass

2. The male gamete is the _____, and the female gamete is the _____.
   a. testis, ovary
   b. testosterone, estrogen
   c. vas deferens, oviduct
   d. sperm, egg

3. Which of the following is not a characteristic of Sertoli cells?
   a. located between the seminiferous tubules
   b. secrete inhibin
   c. stimulated by testosterone
   d. support developing sperm

4. The epididymis:
   a. is a tube connecting the vas deferens to the urethra.
   b. delivers secretions of the seminal vesicles into the urethra.
   c. is a coiled duct just outside the testis.
   d. is a mass of tightly coiled tubules where sperm are produced.

5. Which of the following glands is incorrectly matched with its secretion?
   a. seminal vesicles—testosterone
   b. prostate gland—alkaline fluid
   c. bulbourethral gland—mucus
   d. all of the above are correctly matched

6. In the process of sperm development, secondary spermatocytes develop into _____ which mature to become _____.
   a. primary spermatocytes, spermatids
   b. primary spermatocytes, sperm
   c. spermatids, sperm
   d. spermatids, spermatogonia

7. What part of a sperm contains enzymes that help aid penetration of the egg?
   a. head
   b. acrosome
   c. midpiece
   d. tail

8. Sperm are:
   a. haploid.
   b. diploid.
   c. diploid when they leave the male, but haploid by the time they reach the oviduct.
   d. diploid as they mature in the epididymis, but haploid at the time of fertilization.

9. Sperm production:
   a. is controlled by a positive feedback loop.
   b. occurs in the seminiferous tubules.
   c. is stimulated by inhibin.
   d. begins shortly after birth and continues through old age.

10. The secretion of testosterone is initiated by _____ released by the hypothalamus.
    a. GnRH
    b. LH
    c. FSH
    d. ADH

11. If the pathway from the uterus to the oviduct is blocked:
    a. an oocyte cannot be released.
    b. an oocyte cannot enter the oviduct.
    c. fertilization cannot occur.
    d. the uterine cycle will be disrupted.

12. The _____ provides the muscular force to expel a mature fetus during labor and birth.
    a. fimbriae
    b. endometrium
    c. myometrium
    d. cervix

13. At birth, the ovary:
    a. already contains approximately one million secondary oocytes.
    b. produces high levels of LH to prevent maturation of follicles until puberty.
    c. is controlled by inhibin released by the anterior pituitary.
    d. contains many more immature follicles than will be present at puberty.

14. Which of the following is not released from the ovary with the secondary oocyte?
    a. the polar body
    b. the zona pellucida
    c. the granulosa cells
    d. the corpus luteum

15. The function of the corpus luteum is to:
    a. protect the secondary oocyte from fertilization by multiple sperm.
    b. nourish the secondary ooctye until it reaches the uterus.
    c. secrete progesterone and estrogen to prepare the uterus for implantation.
    d. secrete progesterone and estrogen to prevent release of a second egg before the end of the uterine cycle.

16. Menstruation is triggered by:
    a. the implantation of a fertilized egg.
    b. the process of ovulation.
    c. a decline in the level of progesterone and estrogen.
    d. an increase in the level of progesterone and estrogen.

17. If fertilization does not occur, changing hormone levels leading to menstruation will result from:
    a. the absence of a chorion.
    b. the degeneration of the corpus luteum.
    c. the growth of a new follicle.
    d. proliferation of the endometrium.

18. The first half of the menstrual cycle is influenced by positive feedback. This occurs when:
    a. rising levels of estrogen in the first half of the cycle trigger a surge in LH.
    b. high levels of estrogen and progesterone inhibit LH and FSH.
    c. decreasing levels of estrogen stimulate an increase in LH.
    d. decreasing levels of progesterone stimulate an increase in LH and FSH.

19. Erection occurs when:
    a. arterioles leading into the vascular compartments of the penis dilate, while veins draining the penis are compressed.
    b. arterioles leading into the vascular compartment of the penis constrict, while veins draining the penis dilate.
    c. muscular contractions of the ejaculatory duct compress the vessels that supply the vascular compartments of the penis.
    d. smooth muscle in the penis contracts due to stimulation by the sympathetic nervous system.

20. Hormonal methods of birth control administer progesterone and estrogen in amounts that:
    a. stimulate the release of GnRH.
    b. stimulate the release of LH and FSH.
    c. inhibit the release of LH and FSH.
    d. prevent the proliferative phase of the uterine cycle.

21. One of the drawbacks in using a diaphragm is:
    a. the insertion of the diaphragm must be done in a doctor's office.
    b. a high failure rate.
    c. that diaphragms do not provide protection against STDs.
    d. the side effects, which may include headaches and nausea.

22. Male infertility usually results from:
    a. unusually high levels of testosterone.
    b. failure of the testis to descend completely into the scrotum.
    c. insufficient numbers of normal sperm.
    d. adrenal tumors resulting in increased estrogen secretion.

23. Endometriosis is a condition in which:
    a. the endometrium fails to proliferate during the uterine cycle.
    b. the endometrium is insensitive to estrogen and progesterone.
    c. there is abnormal thickening of the myometrium, creating pressure on the endometrium.
    d. endometrial tissue migrates up the oviduct and grows on other organs.

24. Which of the following types of STDs is incorrectly matched with its causative organism?
    a. gonorrhea—bacteria
    b. Hepatitis B—virus
    c. genital warts—virus
    d. syphilis—virus

25. Genital herpes:
    a. is curable.
    b. is responsible for an increased risk of cervical cancer.
    c. can cause blindness in infants infected during birth.
    d. causes a discharge from the penis or vagina.

Key Concept Review Questions

Each of the Key Concepts listed at the beginning of this chapter has been rewritten as a question below. After successfully completing the study guide exercises and the Chapter Test, you should be able to answer each of these questions. Refer to the Key Concepts list at the beginning of this chapter to check your answers.

1. What are the general structures of a reproductive system?
2. What are the primary organs of the male reproductive system, and what are their functions?
3. Where are sperm produced, and where do they mature?
4. What tubular structures carry the sperm outside the body? What fluid contains the sperm?
5. What is contained in semen?
6. How many chromosomes are there in a sperm cell? What type of cell division produces sperm?
7. What are three hormones important in regulating the male reproductive system?
8. What are the primary organs of the female reproductive system, and what is their function?
9. What is the oviduct, and what is its function? Where does fertilization occur?
10. What is the role of the uterus? Name the two layers of the uterus.
11. What are two functions of the vagina?
12. What is the function of the mammary glands?
13. What are the two organs involved in the menstrual cycle?
14. Which cycle results in the release of a secondary oocyte?
15. Which cycle prepares the endometrium to accept and nurture a fertilized egg?
16. What are five hormones involved in regulating the female reproductive cycle?
17. The human sexual response coordinates sexual function to accomplish what two events?
18. Birth control methods are designed to prevent what general event?
19. What are the four general classes of reproductive disorders?
20. Which STDs are curable and which are not curable? List several STDs of each type.

17

Cell Reproduction and Differentiation

Chapter Summary and Key Concepts

After reading and studying this chapter you should know the following:

Sections 17.1, 17.2

1. The cell cycle consists of two phases: interphase and the mitotic phase.

2. Interphase is a growth phase consisting of three stages: G1, S, and G2. During G1, cells grow; during S, the deoxyribonucleic acid (DNA) is replicated; and during G2, cells continue to grow in preparation for cell division.

3. The mitotic phase consists of a nuclear division called mitosis, and a cytoplasmic division called cytokinesis.

4. Human cells contain DNA arranged into 46 chromosomes; each chromosome includes segments of DNA, called genes, that encode a particular protein.

5. Replication is the process of copying a cell's DNA and occurs during the S phase of the cell cycle.

6. A replicated chromosome consists of two sister chromatids, joined by a centromere.

7. Transcription and translation occur when the information in a DNA gene is used to make a protein.

8. Transcription occurs in the nucleus and makes an RNA copy of a DNA gene.

9. Translation occurs in the cytoplasm and results in production of a protein.

10. Three types of RNA participate in translation: messenger RNA carries instructions about how to build the protein, transfer RNA transports amino acids to the growing protein, and ribosomal RNA forms the ribosome.

11. A mutation is a permanent change in the DNA.

12. Mutations may be helpful or harmful, and if not repaired, they may contribute to the process of evolution.

Sections 17.3, 17.4, 17.5

13. Mitosis can be divided into a series of phases called prophase, metaphase, anaphase, and telophase.

14. Mitosis produces two diploid daughter cells that are identical to the parent cell.

15. Meiosis consists of two successive nuclear divisions called meiosis I and meiosis II.

16. Meiosis produces four haploid cells that are genetically different from the parent cell.

17. In a female, meiosis results in the production of one functional egg. In a male, meiosis results in the production of four functional sperm.

18. Selective gene expression regulates the cell cycle and differentiation.

19. Structural genes encode enzymes or structural proteins; regulatory genes encode repressor or activator proteins.

20. Differentiation is the process whereby cells become different from each other.

Exercises

Complete the exercises for each section after you have read and studied the section. If you cannot answer some questions, or answer them incorrectly, return to the chapter and review this information. You may find it helpful to work on only one section at a time. When you have completed all sections, take the Chapter Test as an indicator of your mastery of this topic.

17.1 The cell cycle creates new cells

17.2 Replication, transcription, and translation—an overview

Matching

____ 1. **cell cycle**   a. a short segment of DNA that codes for a single protein

____ 2. **interphase**   b. a permanent change in the DNA

____ 3. **mitotic phase**   c. an RNA molecule that carries amino acids to the ribosome during translation

____ 4. **deoxyribonucleic acid (DNA)**   d. a base triplet on a tRNA that is complimentary to a mRNA codon

_____ 5. **chromosome**   e. the sequence of events from the time a cell is produced until it completes its own division

_____ 6. **replication**   f. the type of RNA that complexes with protein to make ribosomal subunits

_____ 7. **gene**   g. the process of using an RNA template to produce a protein

_____ 8. **transcription**   h. a type of RNA carries the "message" of a DNA gene specifying the amino acid sequence of a protein

_____ 9. **translation**   i. fragments of DNA that are excluded from the final transcript

_____ 10. **mutation**   j. a group of three bases that encode a specific amino acid

_____ 11. **RNA polymerase**   k. sequences of DNA that carry genetic information and are included in the final transcript

_____ 12. **introns**   l. a structure that consists of rRNA and protein, and joins amino acids together during translation

_____ 13. **exons**   m. the phase of the cell cycle that includes the nuclear and cytoplasmic divisions

_____ 14. **messenger RNA (mRNA)**   n. a DNA molecule associated with histone proteins

_____ 15. **codon**   o. the enzyme that synthesizes RNA

_____ 16. **transfer RNA (tRNA)**   p. a helical, double-stranded string of nucleotides

_____ 17. **anticodon**   q. the longest phase of the cell cycle, consisting of G1, S, and G2

_____ 18. **ribosome**   r. the process of making a complementary RNA copy of a DNA gene

_____ 19. **ribosomal RNA**   s. the process of copying a cell's DNA prior to cell division

Fill-in-the-Blank

20. Division of the cytoplasm is called _____. (17.1)

21. Cells that are not dividing and stop growing during G1 are said to be in a(n) _____ G1 phase. (17.1)

22. In the cell cycle, the cell prepares for division during _____. (17.1)

23. During interphase, chromosomes appear as a grainy material called _____. (17.2)

24. Proteins called _____ combine with DNA and help to maintain the structure of a chromosome. (17.2)

continued

25. During DNA replication, new nucleotides are positioned and linked together by an enzyme called _____ _____. (17.2)

26. Mistakes in base pairing made during DNA replication can often be corrected by DNA _____ enzymes. (17.2)

27. DNA replication occurs simultaneously at many points along a chromosome called replication _____. (17.2)

28. In RNA, the base thymine is replaced by the base _____. (17.2)

29. The beginning of a gene is marked by a unique base sequence called the _____. (17.2)

30. If one strand of a DNA molecule has the base sequence "ATCGGA," the base sequence of the complementary strand will be _____. (17.2)

31. The codon "AUC" codes for the amino acid (abbreviated name) _____. (17.2)

Matching

For each process depicted in Figure 17.1, select four associated terms or characteristics from the list below.

a. transcription
b. translation
c. replication
d. occurs in the cytoplasm
e. occurs in the nucleus
f. occurs in the nucleus
g. converts an mRNA code into a protein
h. produces an exact copy of DNA
i. converts a segment of DNA into RNA
j. sister chromatids, joined by a centromere, are produced
k. ribosomes participate in this process
l. introns and exons are produced

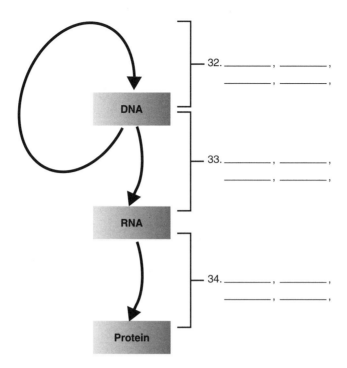

Figure 17.1

Completion

Use the base sequence of the DNA gene below to:

a. write the base sequence of the complementary DNA strand

b. write the base sequence of an mRNA strand produced from this gene

c. write the amino acid sequence of the protein produced from this gene, assuming this base sequence contains no introns

DNA gene   **ACTACCCATGCCAG**

35. Complementary DNA strand: _____

36. Messenger RNA: _____

37. Protein amino acid sequence: _____

Labeling

Use the terms and events of translation listed below to label Figure 17.2.

Terms
a. tRNA
b. start codon
c. amino acids
d. completed protein
e. stop codon
f. ribosomal subunits
g. mRNA
h. anticodon

Events
i. bond forms between amino acids
j. ribosome moves along mRNA
k. tRNA is released
l. tRNA captures free amino acids
m. elongation
n. termination
o. initiation

Figure 17.2

17.3 Cell reproduction—one cell becomes two

17.4 How cell reproduction is regulated

17.5 Environmental factors influence cell differentiation

Matching

____ 1. **mitosis** a. a phase of mitosis when duplicated chromosomes separate and move toward opposite sides of the cell

____ 2. **cytokinesis** b. the process whereby a cell becomes different from its parent or sister cells

____ 3. **prophase** c. a cell that contains both members of each homologous pair

____ 4. **mitotic spindle** d. genes that encode repressor or activator proteins

____ 5. **metaphase** e. division of the nucleus

____ 6. **anaphase** f. cells that have only one member of each homologous pair

____ 7. **telophase** g. duplicated homologous chromosomes pair up and exchange sections of DNA during prophase of meiosis I

____ 8. **diploid cell** h. a cell division process that produces haploid cells

____ 9. **homologous chromosomes** i. division of the cytoplasm

____ 10. **haploid cell** j. genes that encode an enzyme or structural protein

____ 11. **meiosis** k. an arrangement of microtubules that facilitates equal division of the DNA

____ 12. **crossing-over** l. chromosomes that have copies of the same genes in the same locations

____ 13. **regulatory genes** m. phase of mitosis when duplicated chromosomes first become visible

____ 14. **structural genes** n. a phase of mitosis when the two sets of chromosomes arrive at opposite poles of the cell

____ 15. **differentiation** o. a phase of mitosis when duplicated chromosomes align at the center of the cell

Completion

Place an "X" in the column under Mitosis, Meiosis I, or Meiosis II to indicate which cell division process includes the event described.

Event	Mitosis	Meiosis I	Meiosis II
16. crossing-over occurs between homologous chromosomes			
17. produces two haploid cells that still contain sister chromatids			
18. daughter cells are diploid			
19. in human cells 23 duplicated chromosomes will align on the equatorial plate			
20. daughter cells are exactly like the parent cell			
21. produces a secondary spermatocyte in a male			
22. produces an egg in a female			

Labeling

Use the terms and events of mitosis listed below to label Figure 17.3.

Terms
a. cleavage furrow
b. mitotic spindle forming
c. daughter chromosomes
d. centrioles
e. mitotic spindle
f. centromere
g. duplicated chromosome

Events
h. anaphase
i. metaphase
j. telophase and cytokinesis
k. prophase
l. nuclear membrane forms
m. chromosomes uncoiling

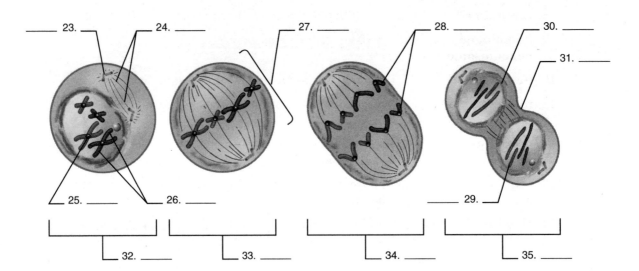

Figure 17.3

Short Answer

A human liver cell is diploid and has 46 chromosomes. Answer the following questions about a liver cell.

36. How many chromosomes will be present in a liver cell in prophase?

37. How does the amount of DNA present in a prophase cell compare to the amount of DNA present in a cell in G1?

38. Does a cell in G2 contain duplicated chromosomes?

39. Is the DNA in a cell during G1 in the form of chromosomes or chromatin?

True/False

_____ 40. A gene is expressed when it is copied during replication.

_____ 41. Repressor proteins prevent gene expression.

_____ 42. Selective gene expression contributes to differentiation.

_____ 43. Most nerve cells divide throughout adulthood.

_____ 44. G1 is one point at which regulation of the cell cycle can occur.

_____ 45. Activator proteins cannot activate a structural gene if a repressor protein is present.

_____ 46. An embryo at the four-cell stage has not yet begun the process of cell differentiation.

_____ 47. Cell differentiation in the early embryo can be triggered by environmental influences.

Chapter Test

Multiple Choice

1. In which phase of the cell cycle does DNA replication occur?
   a. G1
   b. S
   c. G2
   d. M

2. The mitotic phase of the cell cycle consists of:
   a. two nuclear divisions.
   b. two cytoplasmic divisions.
   c. a nuclear division and a cytoplasmic division.
   d. two nuclear divisions and a cytoplasmic division.

3. Histone proteins:
   a. contribute to the structure of a chromosome.
   b. become part of the mitotic spindle.
   c. disappear during prophase.
   d. direct chromosome movement during prophase.

4. A short segment of DNA that represents the code for a specific protein is a:
   a. codon.
   b. triplet.
   c. gene.
   d. chromosome.

5. When a protein has been produced from the information in a molecule of DNA, which of the following has occurred?
   a. replication
   b. transcription
   c. translation
   d. b and c
   e. all of the above

6. The enzyme involved in positioning and joining complementary bases during replication is:
   a. DNA replicase.
   b. DNA transcriptase.
   c. DNA polymerase.
   d. DNA nuclease.

7. Permanent damage to the DNA is referred to as a(n):
   a. mistake.
   b. mutation.
   c. translation.
   d. alteration.

8. Which of the following is a way in which replication and transcription differ?
   a. Only replication requires complementary base pairing.
   b. Only transcription requires an enzyme.
   c. Transcription copies only one side of the DNA molecule.
   d. Replication occurs in the nucleus.

9. A final mRNA transcript will contain:
   a. introns and exons.
   b. introns only.
   c. exons only.
   d. a, b, or c, depending on the needs of the cell at the time.

10. _____ contains codons and _____ contains anticodons.
    a. tRNA, rRNA
    b. mRNA, tRNA
    c. mRNA, rRNA
    d. rRNA, tRNA

11. All proteins in the process of being produced will begin with the amino acid _____.
    a. Phe
    b. Ser
    c. Met
    d. every protein begins with a different amino acid, depending on the DNA message

12. In the process of translation, the instructions about the proper amino acid sequence of a new protein are carried by the:
    a. mRNA.
    b. tRNA.
    c. rRNA.
    d. ribosome.

13. During the phase of translation called elongation, the mRNA passes between _____, catalyzing the binding of _____ to each other.
    a. rRNA and tRNA, proteins
    b. two ribosomal subunits, amino acids
    c. two strands of DNA, nucleotides
    d. a ribosome and a tRNA, proteins

14. Which amino acid sequence is specified by the mRNA base sequence GAAUGCCCGCAC?
    a. glu-cys-pro-his
    b. leu-thr-gly-val
    c. met-pro-ala
    d. tyr-gly-arg

15. If the tubular elements of the cytoskeleton were not disassembled prior to mitosis, which of the following would be affected?
    a. formation of sister chromatids
    b. formation of the mitotic spindle
    c. disassembly of the nuclear membrane
    d. formation of a centromere

16. An original chromosome and its copy:
    a. float singly in the cytoplasm during mitosis.
    b. are joined to each other by a centromere.
    c. are attached down the entire length of the chromosome.
    d. form a chromosome pair.

17. During which phase of mitosis do individual chromosomes arrive at opposite poles of the cell?
    a. metaphase
    b. anaphase
    c. prophase
    d. telophase

18. Which of the following would not occur if the cleavage furrow was unable to form?
    a. DNA replication
    b. division of the centromere
    c. cytokinesis
    d. formation of the mitotic spindle

19. Meiosis in a male produces _____ cells, each of which is _____.
    a. two, haploid
    b. two, diploid
    c. four, haploid
    d. four, diploid
    e. one, haploid

20. Genetic recombination in meiosis is accomplished by a process called:
    a. mitosis.
    b. cleavage.
    c. homologous pairing.
    d. crossing-over.

21. A polar body is first produced:
    a. during meiosis I in a female.
    b. during meiosis II in a female.
    c. during meiosis I in a male.
    d. during meiosis II in a male.

22. A gene is expressed when:
    a. it is encoded by a segment of DNA.
    b. it is edited.
    c. it is replicated.
    d. it is transcribed and translated.

23. Repressor and activator proteins are encoded by:
    a. structural genes.
    b. regulatory genes.
    c. activated genes.
    d. expressed genes.

24. Cells separated from each other after the 16-cell stage fail to produce an entire intact organism. This is because:
    a. They require influential chemicals from neighboring cells to develop.
    b. The trauma of separating cells at this stage prevents further development.
    c. DNA at the 16-cell stage is difficult to activate.
    d. The cells have already begun to differentiate.

25. Most human cells contain _____ chromosomes, while the cells of gametes have _____ chromosomes.
    a. 23, 23
    b. 46, 46
    c. 23, 46
    d. 46, 23

Key Concept Review Questions

Each of the Key Concepts listed at the beginning of this chapter has been rewritten as a question below. After successfully completing the study guide exercises and the Chapter Test, you should be able to answer each of these questions. Refer to the Key Concepts list at the beginning of this chapter to check your answers.

1. What are the two primary phases of the cell cycle?
2. What three stages are included in interphase? What occurs during each stage?
3. What are the two divisions that make up the mitotic phase of the cell cycle?
4. How many chromosomes are found in human cells? What is a gene?
5. What is replication, and in what phase of the cell cycle does it occur?
6. Describe the structure of a replicated chromosome.
7. What two processes must occur for the information in DNA to be used to make a protein?
8. What is transcription and where in the cell does it occur?
9. What is translation and where in the cell does it occur?
10. What are the three types of RNA that participate in translation, and what is the function of each?
11. What is a mutation?
12. Are mutations always harmful? What process requires the presence of mutations?
13. What are the four phases of mitosis?
14. How many daughter cells are produced by mitosis? Are they identical to or different from the parent cell?
15. What are the two successive nuclear divisions of meiosis?
16. What is the number and ploidy (haploid or diploid) of the daughter cells produced by meiosis in a male and in a female?
17. How is the product of meiosis different in males and females?
18. Selective gene expression contributes to the regulation of what two processes?
19. What products are encoded by structural genes and regulatory genes?
20. What is differentiation?

18

Cancer: Uncontrolled Cell Division and Differentiation

Chapter Summary and Key Concepts

After reading and studying this chapter you should know the following:

Sections 18.1, 18.2, 18.3

1. Cancer is a disease of cell division and differentiation.

2. Normal cells have carefully regulated cell division, and they generally stay in one location in the body.

3. Cancer cells exhibit unregulated cell division, and may move to distant locations in the body.

4. A tumor is a mass of cells that has lost control over cell division. Tumors may be benign or cancerous.

5. Benign tumors remain in one place as a well-defined mass of cells.

6. A tumor is considered a cancer when the cells lose their organization, structure, and regulatory control.

7. Cancerous tumors that remain in one place are in situ cancers.

8. Malignant tumors are formed by cancer cells that invade normal tissues and metastasize to distant sites.

9. In order for cancer to develop, cells must divide uncontrollably and break away from surrounding cells.

10. Proto-oncogenes are regulatory genes that promote normal cell growth.

11. Tumor suppressor genes are normal regulatory genes that suppress cell growth.

12. Oncogenes are genes that cause cancer and develop from mutated proto-oncogenes.

13. Mutated or damaged proto-oncogenes and tumor suppressor genes contribute to the development of cancer by causing or allowing abnormal cell division.

14. Substances that cause cancer are called carcinogens, and they include viruses and bacteria, environmental chemicals, tobacco, radiation, dietary factors, and alcohol.

15. Susceptibility to cancer can be inherited if parents pass oncogenes to their offspring.

16. Suppression of the immune system may allow some cancers to develop more easily.

Sections 18.4, 18.5

17. Early detection is crucial in treating cancer successfully.

18. Traditional cancer treatments include surgery, radiation, chemotherapy, or a combination of two or more treatments.

Sections 18.6, 18.7

19. The 10 most common cancers are lymphomas, leukemia, and cancers of the skin, breast, prostate, lung, colon and rectum, urinary bladder, uterus, and mouth and pharynx.

20. Most cancers can be prevented by making wise lifestyle and nutritional choices.

Exercises

Complete the exercises for each section after you have read and studied the section. If you cannot answer some questions, or answer them incorrectly, return to the chapter and review this information. You may find it helpful to work on only one section at a time. When you have completed all sections, take the Chapter Test as an indicator of your mastery of this topic.

18.1 Tumors can be benign or cancerous

18.2 Cancerous cells lose control over cell functions

18.3 How cancer develops

Matching

_____ 1. **hyperplasia**   a. the movement of cancer cells away from the main tumor

_____ 2. **tumor**   b. a substance or physical factor that causes cancer

_____ 3. **dysplasia**   c. a mutated proto-oncogene that contributes to the development of cancer

_____ 4. **cancer**   d. the process of transforming a normal cell into a cancerous cell

_____ 5. **in situ cancer**   e. a condition that results from an abnormally high frequency of cell division

_____ 6. **metastasis**   f. a type of cancer involving the pigment-producing cells of the skin

_____ 7. **malignant tumor**   g. a mass of cells that has lost normal regulatory mechanisms over cell division; can be benign or cancerous

_____ 8. **proto-oncogene**   h. a normal regulatory gene that represses cell growth, division, differentiation, or adhesion

_____ 9. **oncogene**   i. an unhealthy or abnormal change in cell structure

_____ 10. **tumor suppressor gene**   j. a general term for a tumor that contains some cells that have lost organization, structure, and regulatory control

_____ 11. **carcinogenesis**   k. a normal regulatory gene that promotes cell growth, differentiation, division, or adhesion

_____ 12. **carcinogen**   l. a cancerous tumor that remains in one place

_____ 13. **melanoma**   m. cancers that invade normal tissue and metastasize to different sites

Fill-in-the-Blank

Referenced sections are in parentheses.

14. _____ is often a sign that tumors are precancerous. (18.1)

15. One in _____ people will die of cancer. (18.2)

16. _____ factors are proteins encoded by regulatory genes, and secreted by cells, that influence nearby cells. (18.3)

17. Cancerous cells display surface _____ that identify them as abnormal to the immune system. (18.3)

18. The development of cancer requires the presence of _____ oncogenes. (18.3)

19. The primary role of the gene p53 is to _____ cell division in cells that appear to be cancerous. (18.3)

20. The _____ virus may contribute to liver cancer. (18.3)

21. Chemical carcinogens that directly damage DNA are called _____. (18.3)

22. Chemicals that increase the potency of a carcinogen are called _____. (18.3)

23. Tobacco accounts for more than _____ % of all cancer deaths. (18.3)

24. Dietary factors are probably involved in up to _____ % of all cancers. (18.3)

25. _____ _____ are reactive fragments of molecules produced during cellular metabolism that can damage healthy molecules in the body. (18.3)

26. Vitamins _____, _____, and _____ appear to have antioxidant benefits. (18.3)

27. Suppression of the _____ system may allow certain cancers to develop more easily. (18.3)

Labeling

Use the terms and descriptions to label Figure 18.1. Some structures will have two terms or descriptions that apply.

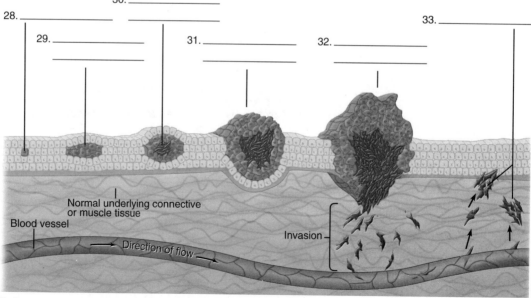

Figure 18.1

a. dysplasia
b. malignant tumor
c. hyperplasia
d. metastases
e. genetically altered epithelial cell

f. in situ cancer
g. cells stay in one place
h. cells change form
i. cancer cells invade normal tissue
j. cells divide more rapidly than normal

18.4 Advances in diagnosis enable early detection

18.5 Cancer treatments

Completion

Name each diagnostic method and cancer treatment described in the table.

Diagnostic Method or Treatment	Description
1.	a diagnostic method in which radioactive substances are used to create three-dimensional images showing the metabolic activity of body structures
2.	a diagnostic method which detects the presence of oncogenes
3.	a diagnostic method in which body fluids are screened for the presence of telomerase
4.	a cancer treatment involving the administration of cytotoxic chemicals to destroy cancer cells
5.	a cancer treatment in which the cancer is targeted with lasers
6.	a potential cancer treatment in which defective genes are repaired or replaced
7.	a cancer treatment involving the use of anti-angiogenic drugs

Short Answer

8. Why are chemicals that inhibit cell division able to target primarily cancer cells while sparing most healthy cells?

9. How can the antigen of a cancer cell be used to develop immunotherapy treatments?

18.6 The 10 most common cancers

18.7 Most cancers can be prevented

Short Answer

Write the symptoms and risk factors for each cancer listed below. Answers for exercises in this section are not in the answer key; refer to the textbook to check your answers.

1. Skin cancer:

2. Breast cancer:

3. Prostate cancer:

4. Lung cancer:

5. Colon and rectal cancer:

6. Lymphoma:

7. Urinary bladder cancer:

8. Uterine cancer:

9. Leukemia:

10. Cancer of the mouth and pharynx:

11. Using the recommendations in the textbook for cancer prevention, list at least six things you can do to reduce your risk of cancer:

    1. _____

    2. _____

    3. _____

    4. _____

    5. _____

    6. _____

Chapter Test

Multiple Choice

1. Cell division in healthy cells is normally inhibited by:
    a. growth factors.
    b. structural genes.
    c. signals from nearby cells.
    d. the immune system.

2. Loss of regulatory control over cell division causes a cell to:
    a. grow unusually large.
    b. secrete inhibiting factors.
    c. divide more frequently than normal.
    d. all of the above.

3. Benign tumors threaten health only if they:
    a. secrete destructive enzymes.
    b. grow large enough to crowd normal cells.
    c. become cancerous.
    d. metastasize.

4. A tumor is defined as cancer when:
    a. it metastasizes.
    b. the cells have lost regulatory control over cell division.
    c. the cells have undergone structural changes.
    d. some of the cells have lost regulatory control over cell division and undergone structural changes.

5. Select the sequence of developments that occurs in forming a malignant tumor, beginning with a single cell.
   a. single mutated cell, dysplasia, hyperplasia, malignant tumor, in situ cancer
   b. single mutated cell, hyperplasia, dysplasia, in situ cancer, malignant tumor
   c. single mutated cell, in situ cancer, dysplasia, hyperplasia, malignant tumor
   d. single mutated cell, dysplasia, in situ cancer, hyperplasia, malignant tumor

6. Which of the following is the same in an in situ tumor and a malignant tumor?
   a. cell structure
   b. cell function
   c. frequency of cell division
   d. tumor organization

7. New colonies of cancer established by cancer cells that have migrated away from the original tumor are called:
   a. in situ tumors.
   b. metastases.
   c. hyperplasias.
   d. migration masses.

8. Cancer ranks _____ as a cause of death.
   a. first
   b. second
   c. third
   d. fourth

9. Tumor suppresor genes are:
   a. normal regulatory genes that promote cell division.
   b. precancerous genes that have mutated at least once.
   c. normal regulatory genes that repress cell growth.
   d. regulatory genes that produce proteins that inhibit oncogenes.

10. Which of the following is *not* true of the gene p53?
    a. p53 encodes a gene that prevents damaged cells from dividing
    b. p53 is a tumor suppressor gene
    c. individuals who carry the p53 gene have an increased risk of cancer
    d. a mutated p53 gene has been found in some cervical cancers

11. A cell is dividing under circumstances when cell division would normally be inhibited. Which of the following is a regulatory gene that has probably been inactivated in the cell?
    a. a proto-oncogene
    b. an oncogene
    c. a tumor suppressor gene
    d. a growth factor

12. The difference between an initiator and a promotor is:
    a. an initiator is a chemical carcinogen, while a promotor is a biological carcinogen.
    b. a promotor is a cell that aids the process of viral infection, while an initiator is a virus that mutates host cell DNA.
    c. an initiator is a chemical carcinogen, while a promotor is a chemical that enhances the effects of carcinogens.
    d. a promotor is an environmental factor that can cause DNA mutation, while an initiator is a mutation that leads to carcinogenesis.

13. One mechanism by which viral infections may increase the risk of cancer is:
    a. viral infections increase the activity of tumor suppressor genes.
    b. the viral DNA sequence alters the function of a proto-oncogene.
    c. viral proteins interfere with maintenance of the plasma membrane.
    d. the virus interferes with the cell's ability to display "self" markers on the cell surface.

14. Sources of radiation that can increase cancer risk include:
    a. the sun.
    b. medical and dental x-rays.
    c. radon gas.
    d. all of the above.

15. Diet and tobacco together account for approximately _____ of all cancers.
    a. 10%
    b. 40%
    c. 60%
    d. 90%

16. Inheriting a mutated proto-oncogene from one or both of your parents means that:
    a. you will get cancer at a younger age than normal.
    b. you will get cancer, although it may be at an older age.
    c. you are at an increased risk for cancer.
    d. your risk for cancer is no different from anyone else's.

17. Which of the following statements about sunlight and skin cancer is true?
    a. Sunlamps and tanning booths are not as dangerous as natural sunlight.
    b. Sun exposure is only dangerous when it results in a sunburn.
    c. Any cumulative exposure to the sun increases the risk of cancer.
    d. Sunburns experienced during childhood do not increase the risk of cancer.

18. The presence of free radicals may increase the risk of cancer because:
    a. free radicals reduce a cell's ability to fight viral infections.
    b. free radicals increase sun damage to cells.
    c. free radicals can damage a cell's DNA.
    d. free radicals interfere with normal DNA repair mechanisms.

19. As cells become cancerous:
    a. the nucleus becomes smaller.
    b. the amount of cytoplasm increases.
    c. they lose their specialized structures.
    d. all of the above.

20. In order for cancer to develop, what two things must happen?
    a. both a tumor suppressor gene and an oncogene must be mutated
    b. cells must divide uncontrollably, and they must experience physical changes that allow them to break away from surrounding cells
    c. dietary and environmental factors must be present that act as carcinogens
    d. a benign tumor must form, followed by hyperplasia

21. An enzyme that occurs rarely in normal cells but frequently in cancer cells is:
    a. prolactase.
    b. mutase.
    c. DNAse.
    d. telomerase.

22. The side effects of chemotherapy include nausea. This is because:
    a. the toxins in the drugs are difficult for the body to metabolize.
    b. rapidly dividing healthy cells, such as occur in the lining of the digestive tract, are also destroyed.
    c. chemotherapy interferes with the normal absorption of nutrients from the digestive tract.
    d. as the chemicals are metabolized, the by-products increase acid production in the stomach.

23. The growth of new blood vessels to support rapidly growing tumors is called:
    a. angiogenesis.
    b. cardiovascular supplementation.
    c. vascularization.
    d. capillary regeneration.

24. Which of the following skin cancers is incorrectly matched with its characteristic?
    a. basal call cancer—metastasizes quickly
    b. squamous cell cancer—arises from epithelial cells
    c. melanoma—the least common form of skin cancer
    d. melanoma—the deadliest form of skin cancer

25. Cancer of the blood-forming organs:
    a. is called lymphoma.
    b. occurs more often in children than adults.
    c. is usually treated by radiation.
    d. may be caused by a virus.

Key Concept Review Questions

Each of the Key Concepts listed at the beginning of this chapter has been rewritten as a question below. After successfully completing the study guide exercises and the Chapter Test, you should be able to answer each of these questions. Refer to the Key Concepts list at the beginning of this chapter to check your answers.

1. Cancer is a disease involving what two cellular processes?
2. What are the two primary characteristics of normal cells that are altered in cancer cells?
3. What are the two defining characteristics of cancer cells?
4. What characterizes the cell division of cells in a tumor? Are tumors always cancerous?
5. Describe a benign tumor.
6. What three things must tumor cells lose to be considered cancerous?
7. What is an in situ cancer tumor?
8. What is a malignant tumor?
9. What two things must happen in order for cancer to develop?
10. What are proto-oncogenes?
11. What are tumor suppressor genes?

12. What are oncogenes? How do oncogenes develop?
13. How do mutated proto-oncogenes or tumor suppressor genes contribute to the development of cancer?
14. What is a carcinogen? Name some common carcinogens.
15. How can susceptibility to cancer be inherited?
16. When does the immune system contribute to the development of cancer?
17. What is crucial in treating cancer successfully?
18. What are four traditional cancer treatments?
19. What are the 10 most common cancers?
20. Is cancer preventable?

19

Genetics and Inheritance

Chapter Summary and Key Concepts

After reading and studying this chapter you should know the following:

Sections 19.1, 19.2

1. Genetics is the study of genes and their inheritance.

2. Human cells contain 22 pairs of chromosomes called autosomes and one pair of sex chromosomes.

3. Alleles are structural and functional variations of homologous genes.

4. An individual is homozygous when he or she possesses two identical alleles, and heterozygous when he or she possesses two different alleles.

5. An individual's complete set of genes and alleles is the genotype, and his or her observable physical and functional traits is the phenotype.

6. A Punnett square is a method for determining patterns of inheritance.

7. Gregor Mendel proposed the basic rules of inheritance in the 1850s.

8. The law of segregation states that gametes receive only one allele of each gene pair.

9. The law of independent assortment states that genes for different traits are separated from each other during meiosis.

10. In complete dominance, a dominant allele masks the effect of a recessive allele.

11. In incomplete dominance, the heterozygous genotype produces a phenotype that is intermediate between the two homozygous phenotypes.

12. In codominance, both alleles present will be equally expressed.

13. In polygenic inheritance, a phenotypic trait depends on the action of multiple genes.

14. Many phenotypes are determined by genotype and environmental effects.
15. Linked genes are genes located on the same chromosome and may be inherited together.

Sections 19.3, 19.4, 19.5, 19.6

16. A non-sex gene located on a sex chromosome shows a pattern of sex-linked inheritance.
17. A trait that is not carried on a sex chromosome but is influenced by the expression of a sex chromosome is a sex-influenced trait.
18. Chromosome structure or number may be altered by nondisjuction, deletions, and translocations.
19. Nondisjunction occurs when homologous chromosomes or sister chromatids fail to separate during cell division, a deletion occurs when a piece of a chromosome breaks off and is lost, and a translocation occurs when a piece of a chromosome breaks off and reattaches at another site.
20. Inherited genetic disorders usually involve recessive alleles.

Exercises

Complete the exercises for each section after you have read and studied the section. If you cannot answer some questions, or answer them incorrectly, return to the chapter and review this information. You may find it helpful to work on only one section at a time. When you have completed all sections, take the Chapter Test as an indicator of your mastery of this topic.

19.1 Alleles are different forms of homologous genes

19.2 Genetic inheritance follows certain patterns

Matching

____ 1. **genetics** a. a heritable change in the DNA

____ 2. **inheritance** b. a type of inheritance in which the presence of one dominant allele is sufficient to cause expression of the dominant phenotype

____ 3. **autosome** c. genes that are on the same chromosome

____ 4. **sex chromosomes** d. an allele that masks the expression of the partner allele on the homologous chromosome

____ 5. **allele** e. the study of genes

____ 6. **homozygous** f. the observable physical and functional traits of an individual

____ 7. **heterozygous** g. an allele that must be present on both homologous chromosomes to be expressed

_____ 8. **mutation**   h. a method used to predict patterns of inheritance

_____ 9. **genotype**   i. a type of inheritance in which the heterozygous genotype results in a phenotype intermediate between the two homozygous phenotypes

_____ 10. **phenotype**   j. a type of inheritance in which the expression of a phenotypic trait depends on many genes

_____ 11. **Punnett square method**   k. chromosomes that determine gender

_____ 12. **dominant allele**   l. something received from an ancestor

_____ 13. **recessive allele** m. a condition in which both alleles for a particular gene are identical

_____ 14. **complete dominance**   n. a chromosome that does not determine gender

_____ 15. **incomplete dominance** o. the complete set of genes and alleles possessed by an individual

_____ 16. **codominance**   p. a structural and functional variation of a homologous gene

_____ 17. **polygenic inheritance** q. a condition in which the alleles for a particular gene are different

_____ 18. **linked genes**   r. a type of inheritance in which both alleles present are fully and equally expressed

Fill-in-the-Blank

Referenced sections are in parentheses.

19. The complete set of genes on the chromosomes of a particular organism is called the _____. (19.1)

20. The basic rules of inheritance were first proposed by an Austrian monk named _____ _____. (19.2)

21. The law of _____ _____ states that genes for different traits are separated from each other during the formation of gametes. (19.2)

22. Phenotype is affected by both the _____ and the _____. (19.2)

23. Genes that are close together on a chromosome and are inherited together are called _____ _____. (19.2)

24. The "factors" of heredity proposed by Mendel are now known to be _____. (19.2)

Labeling

Write in the genotypes of the male and female parent in Figure 19.1, and then complete the Punnett square.

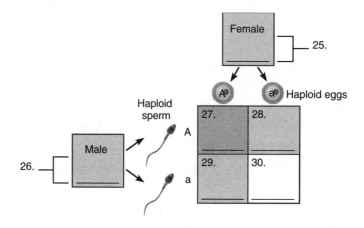

Figure 19.1

Short Answer

31. Tongue-rolling is inherited as a completely dominant trait. Individuals who inherit the dominant allele (T) can roll their tongue, while those who inherit only the recessive allele (t) cannot.
    a. What is the genotype of a heterozygous individual?
    b. What is the genotype of a person who cannot roll their tongue?
    c. What is the genotype of a person who is homozygous dominant?
    d. Use a Punnett square to determine the possible genotypes of all children born to a heterozygous father and a non-tongue rolling mother:

    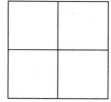

    e. What percentage of the children will be expected to be tongue-rollers?

32. In the inheritance of hair type, homozygous dominant (WW) individuals have curly hair, homozygous recessive (ww) individuals have straight hair, and heterozygous (Ww) individuals have wavy hair.
    a. What type of inheritance is occurring?
    b. Use a Punnett square to determine the genotypes of the children born to a curly-haired father and a straight-haired mother.

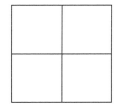

c. What percentage of the children will have wavy hair?

33. The inheritance of the ABO blood type is an example of both complete dominance and codominance. Alleles A and B are codominant, and allele O is recessive.
   a. What is the genotype of an individual with type O blood?
   b. What is the blood type of an individual who carries both the A allele and the B allele?
   c. Use a Punnett square to represent the possible genotypes of the children born to a man with blood type AB, and a woman with blood type O.

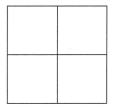

19.3 Sex-linked inheritance: X and Y chromosomes carry different genes

19.4 Chromosomes may be altered in number or structure

19.5 Inherited genetic disorders usually involve recessive alleles

19.6 Genes encode for proteins, not for specific behaviors

Fill-in-the-Blank

Referenced sections are in parentheses.

1. A composite display of all of the chromosomes of an organism is called a _____. (19.3)

2. Human cells contain 22 pairs of chromosomes called _____ and one pair of _____ chromosomes. (19.3)

3. In humans, maleness is determined by the presence of a _____ chromosome. (19.3)

continued

4. An inheritance pattern that depends on genes located on sex chromosomes is called _____-_____ inheritance. (19.3)

5. A trait that is not inherited on a sex chromosome but is influenced by the expression of a sex chromosome is called a _____-_____ trait. (19.3)

6. A _____ occurs when a piece of a chromosome breaks off and is lost. (19.4)

7. A _____ occurs when a piece of a chromosome breaks off and reattaches at another site. (19.4)

8. Huntington's disease is an example of a _____ lethal allele. (19.5)

Short Answer

9. Why do more males than females have X-linked disorders?

10. Why can a man inherit an X-linked trait only from his mother?

11. Why will all the daughters of a man with an X-linked trait be carriers for the disorder?

12. Color blindness is inherited as an X-linked recessive trait. Use X^C to represent the allele for normal vision, and X^c to represent the allele for color-blindness.
    a. What is the genotype of a color-blind woman?
    b. What is the genotype of a woman with normal vision who is a carrier?
    c. What is the genotype of a color-blind man?
    d. Use a Punnett square to determine the possible genotypes of children born to a man with normal vision and a woman who is color-blind.

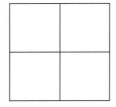

e. What percentage of their sons will be color-blind?

f. What percentage of their daughters will be color-blind?

13. Why might the ability of genes to encode proteins influence behavior and personality?

Completion

Identify each genetic disorder described below.

Genetic Disorder	Description
14.	Individuals with this disorder have only one X chromosome, are sterile females, and may have shortened life expectancy.
15.	This disorder is inherited as an autosomal recessive and develops because affected individuals are unable to metabolize phenylalanine.
16.	This disorder is caused by a deletion in chromosome 5, and affected infants are usually mentally and physically retarded, with a cat-like cry.
17.	Affected men have the genotype XXY, are tall and sterile, and may have mild mental impairment.
18.	This disorder is most often caused by an extra copy of chromosome 21.

Labeling

Draw the chromosome in the cells below indicating the results of nondisjunction at different stages of meiosis. (Refer to Figure 19.13 of the textbook to check your answers.)

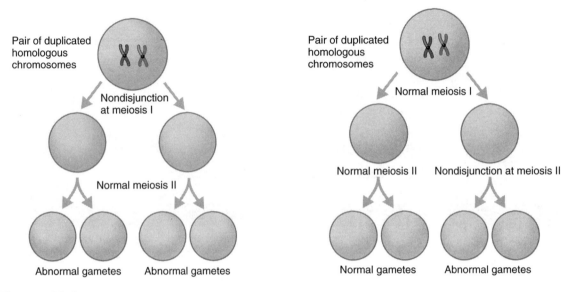

Figure 19.2

Chapter Test

Multiple Choice

1. Human cells contain 22 pairs of _____.
   a. chromosomes
   b. alleles
   c. homologues
   d. autosomes

2. An individual whose genotype is Ww is referred to as being:
   a. homozygous.
   b. heterozygous.
   c. genotypic.
   d. dominant.

3. The law of segregation states that:
   a. in the formation of gametes, alleles separate from each other so that each gamete receives only one allele.
   b. male and female gametes segregate independently of each other.
   c. the genes on one chromosome move independently of the genes on another chromosome.
   d. only alleles that are not linked will be segregated during the formation of gametes.

4. A recessive phenotype will be expressed when:
   a. at least one recessive allele is present.
   b. at least one dominant allele is present.
   c. both alleles present are recessive.
   d. both alleles present are dominant.

5. In complete dominance, the recessive allele usually:
   a. codes for only a small amount of a protein.
   b. codes for a fully functional protein.
   c. codes for a nonfunctional protein.
   d. codes for a protein that is repressed by the dominant allele.

6. When a white horse and a chestnut horse produce a palomino horse, this is an example of:
   a. complete dominance.
   b. incomplete dominance.
   c. codominance.
   d. polygenic inheritance.

7. In the human ABO blood-typing system, an individual with A and B alleles will express both A and B phenotypes on their red blood cells. This is an example of:
   a. complete dominance.
   b. incomplete dominance.
   c. codominance.
   d. polygenic inheritance.

8. In humans, the allele for a widow's peak (W) is dominant. Which of the following represent the possible phenotypes in the offspring of two parents, both with a straight hairline?
   a. All offspring will have a widow's peak.
   b. All offspring will have a straight hairline.
   c. Offspring will show both phenotypes.
   d. Most of the offspring will have a straight hairline, but it is possible to have a widow's peak.

9. When continuous variation occurs in a phenotypic trait, the trait is likely due to:
   a. complete dominance.
   b. incomplete dominance.
   c. codominance.
   d. polygenic inheritance.

10. The frequency with which genes on a particular chromosome are inherited together depends on:
    a. whether they are dominant or recessive.
    b. the occurrence of nondisjunction during meiosis.
    c. how close together they are on the chromosome.
    d. the structure of the mitotic spindle.

11. In humans, femaleness is determined by:
    a. the presence of at least one X chromosome.
    b. the presence of two X chromosomes.
    c. the absence of a Y chromosome.
    d. the presence of adequate levels of estrogen.

12. Testosterone is responsible for the development of the male reproductive system, and it is produced by the testis. The development of the testes depends on the presence of a Y chromosome. What might occur if the Y chromosome is present, but the testes are unable to produce testosterone?
    a. normal male development
    b. normal female development
    c. the testes would be present, but further male development would be arrested
    d. testes and ovaries would be present

13. Hemophilia is an X-linked recessive trait resulting in the inability to clot the blood. A man with hemophilia inherited the condition from:
    a. his mother.
    b. his father.
    c. both of his parents.
    d. all of the above could produce hemophilia in a man.

14. A woman with hemophilia inherited the condition from:
    a. her mother.
    b. her father.
    c. both of her parents.
    d. all of the above could produce hemophilia in a woman.

15. Baldness is a sex-influenced trait. Women must inherit both recessive alleles to be bald, while men need only inherit one recessive allele. This occurs because:
    a. estrogen masks the effect of the baldness allele.
    b. the presence of two X chromosomes in a woman masks the baldness allele.
    c. the baldness allele is linked to the Y chromosome.
    d. testosterone stimulates the expression of the baldness allele.

16. Nondisjunction in meiosis II results in failure of:
    a. homologous chromosomes to separate.
    b. sister chromatids to separate.
    c. gametes to separate.
    d. centromeres to separate.

17. An infant born with a deletion in chromosome 5 will have:
    a. Down syndrome.
    b. Turner syndrome.
    c. Edwards' syndrome.
    d. cri du chat syndrome.

18. An infant born with the genotype XYY will:
    a. be phenotypically female and fertile.
    b. be phenotypically male, short, and mildly retarded.
    c. be phenotypically male and usually normal.
    d. show partial male and female development.

19. Dominant alleles that cause disease are especially dangerous because:
    a. the severity of the disease is greater than those caused by recessive alleles.
    b. heterozygous individuals will develop the disease.
    c. individuals must inherit two dominant alleles to show the disease.
    d. dominant alleles are inherited more frequently than recessive alleles.

20. Different alleles probably arose through:
    a. mutations.
    b. recombinant DNA technology.
    c. genetic engineering.
    d. inheritance patterns.

21. Alterations in chromosomal structure or number that occur during mitosis or meiosis are seldom seen because:
    a. they do not affect the organism.
    b. the effects are so mild that they often go unrecognized.
    c. the effects are usually masked by more serious conditions.
    d. the altered genes are crucial to development and the organism dies.

22. A man carries a Y-linked trait. He will pass this trait to:
    a. all of his daughters.
    b. all of his sons.
    c. 50% of his daughters and 100% of his sons.
    d. 50% of his sons and none of his daughters.

23. When a translocation occurs:
    a. a piece of a chromosome breaks off and is lost.
    b. a damaged chromosome is passed from parent to child.
    c. an X-linked trait is passed to a son.
    d. a segment of a chromosome breaks off and reattaches at another location.

24. Which of the following is a condition caused by a dominant-lethal allele?
    a. Down syndrome
    b. Huntington disease
    c. Klinefelter syndrome
    d. Hemophilia

25. Specific genes code for specific:
    a. proteins.
    b. behaviors.
    c. traits.
    d. allele combinations.

Key Concept Review Questions

Each of the Key Concepts listed at the beginning of this chapter has been rewritten as a question. After successfully completing the study guide exercises and the Chapter Test, you should be able to answer each of these questions. Refer to the Key Concepts list at the beginning of this chapter to check your answers.

1. What is genetics?
2. How many pairs of chromosomes, and of what type, do human cells contain?
3. What are alleles?
4. What makes an individual homozygous or heterozygous?
5. How would you define genotype and phenotype?
6. What is a Punnett square used for?
7. Who proposed the basic rules of inheritance, and in what year?
8. What principle is stated by the law of segregation?
9. What principle is stated by the law of independent assortment?
10. What occurs in complete dominance?
11. How is incomplete dominance different from complete dominance?
12. How are alleles expressed in codominance?
13. What determines phenotype in polygenic inheritance?

14. What, besides genotype, can influence the phenotypic expression of a trait?

15. What are linked genes?

16. What causes sex-linked inheritance?

17. What is a sex-influenced trait?

18. What are three ways in which chromosome number or structure can be altered?

19. What occurs in nondisjunction? What is a deletion? What is a translocation?

20. Do inherited disorders usually involve recessive or dominant alleles?

20

DNA Technology and Genetic Engineering

Chapter Summary and Key Concepts

After reading and studying this chapter you should know the following:

Sections 20.1, 20.2

1. Biotechnology is the technical application of biological knowledge for human purposes.

2. DNA sequencing allows scientists to determine the base sequence of a DNA molecule.

3. DNA sequencing relies on complementary base pairing to produce a new DNA strand, and requires the use of short single-stranded DNA segments, primers, modified DNA nucleotides, DNA polymerase, gel electrophoresis, and a fluoroscope.

4. Recombinant DNA is produced by combining DNA from two different sources. Recombinant DNA technology provides tools for manipulating DNA, and involves cutting, splicing, and copying DNA.

5. Recombinant bacteria are produced using DNA from a foreign organism, restriction enzymes, DNA ligases, and plasmids.

6. Restriction enzymes cut DNA by breaking the bonds between specific base pairs.

7. DNA ligases can bind DNA fragments from different organisms together after they have been cut by restriction enzymes.

8. Plasmids are extrachromosomal DNA molecules that occur naturally in bacteria.

9. Plasmids can accept a foreign piece of DNA and carry it back into the bacteria. Bacteria containing a foreign gene can manufacture a foreign protein.

10. Polymerase chain reaction (PCR) is a technique that amplifies a small fragment of DNA.

11. PCR uses separated strands of a small piece of DNA to serve as a template for the construction of new complementary strands. Cycles of heating and cooling allow the process to repeat until many identical DNA molecules are produced.

Sections 20.3, 20.4

12. Transgenic organisms carry foreign genes as a result of genetic engineering.

13. Transgenic bacteria created with engineered plasmids can produce foreign proteins, including human proteins.

14. Transgenic plants carry foreign genes and have been engineered to resist adverse weather and pests, carry vitamin precursors, and produce vaccines.

15. Transgenic plants have created some concerns about potential safety and environmental hazards associated with their use.

16. Creating transgenic animals is difficult, but research continues because of the great potential benefits.

17. Current techniques for creating transgenic animals include inserting DNA into a single isolated cell, which can give rise to a whole organism.

18. Gene therapy involves the insertion of human genes into human cells for medical purposes.

19. Gene therapy may utilize retroviruses as vectors—agents that can transport a gene into a human cell.

20. Gene therapy has been successfully used to treat Severe Combined Immune Deficiency, and research continues in the treatment of cystic fibrosis and cancer.

Exercises

Complete the exercises for each section after you have read and studied the section. If you cannot answer some questions, or answer them incorrectly, return to the chapter and review this information. You may find it helpful to work on only one section at a time. When you have completed all sections, take the Chapter Test as an indicator of your mastery of this topic.

20.1 DNA sequencing reveals structure of DNA

20.2 DNA can be modified in the laboratory

Matching

____ 1. **biotechnology** a. enzymes that bind DNA fragments together

____ 2. **recombinant DNA technology** b. enzymes that cut DNA into fragments by breaking the bonds between specific base pairs

____ 3. **genetic engineering** c. short double-stranded pieces of DNA that bind to a single strand of DNA and serve as a starting point for the synthesis of a new strand

____ 4. **primers** d. a process that can amplify a small sample of DNA

_____ 5. **DNA polymerase**   e. a process that separates DNA fragments according to size

_____ 6. **gel electrophoresis**   f. the technical application of biological knowledge for human purposes

_____ 7. **restriction enzymes**   g. small, circular, self-replicating DNA molecules found in bacteria

_____ 8. **DNA ligases**   h. an enzyme required for DNA synthesis

_____ 9. **plasmids**   i. tools that allow cutting, splicing, and copying DNA

_____ 10. **polymerase chain reaction**   j. manipulation of the genetic makeup of cells or whole organisms

Completion

Use the terms below to complete the paragraph, then repeat this exercise with the terms covered.

smaller	strands	test tube
fluorescent	labeled	fluoroscope
labeled	synthesis	base
polymerase	synthesis	lengths
primers	size	larger
complementary	original	gel electrophoresis

The process of DNA sequencing allows biologists to determine the (11) _____ sequence of a strand of DNA. First, many copies of a short, single-stranded piece of DNA are placed in a (12) _____. Next, (13) _____ are added. Primers are short, double-stranded pieces of DNA that bind to one end of the (14) _____. A primer is needed to provide a starting point for the (15) _____ of a new complementary DNA strand. Also required are free, available DNA nucleotides, containing the bases A, T, C, and G. A mixture of normal nucleotides and (16) _____ nucleotides will be added to the tube. The labeled nucleotides are identifiable because they have a (17) _____ marker. Labeled nucleotides have also been altered so that DNA (18) _____ will stop as soon as they are added to the chain. Finally, the enzyme DNA (19) _____ will be added; this enzyme is required for DNA synthesis. When the process is completed, many DNA strands of different (20) _____ will have been produced, each ending with a (21) _____ nucleotide. A process called (22) _____ sorts the strands by

continued

(23) _____. As the strands move through the gel column, the (24) _____ pieces move more quickly than the (25) _____ pieces. By stopping the process before the smallest pieces have moved all the way through the gel, the strands will be arranged sequentially from smallest to largest. The bases are identified by scanning the gel column in a (26) _____, a device that measures fluorescence. Because the strands are (27) _____ to the original piece of DNA to be sequenced, determining the base of each terminal nucleotide will provide the base sequence of the (28) _____ DNA strand.

Fill-in-the-Blank

Referenced sections are in parentheses.

29. Restriction enzymes cut DNA between specific _____ sequences at locations called _____ sites. (20.2)

30. A plasmid is an extrachromosomal, self-replicating DNA molecule found in _____. (20.2)

31. Plasmids contain genes needed for bacterial _____. (20.2)

32. Recombinant DNA plasmids contain _____ from a foreign source. (20.2)

33. PCR is not used to clone functional genes because the resulting DNA copies do not contain _____ genes required for the correct gene expression. (20.2)

34. In PCR, the two strands of the DNA molecule to be copied are separated by gentle _____. (20.2)

35. One heating and cooling cycle in PCR will produce _____ DNA molecules from one original DNA molecule. (20.2)

36. Bacteria can manufacture human _____ when they express human genes carried in a recombinant plasmid. (20.2)

37. In the polymerase chain reaction (PCR), the gentle heating is used to _____ a DNA molecule. (20.2)

Labeling

Label the diagram of recombinant DNA production below with the appropriate term or description.

a. DNAs are mixed. Human fragments line up with plasmid by base pairing of exposed single-strand regions.

b. Plasmids are absorbed by bacteria.

c. Human DNA fragments

d. DNA from bacterial and human cells are isolated.

e. DNA ligase is added, to connect human and plasmid DNA together.

f. Both human DNA and plasmid DNA are cut with the same restriction enzyme.

g. Human DNA containing a gene of interest

h. Bacteria containing the recombinant plasmids of interest are selected and cloned.

i. Bacterial plasmid

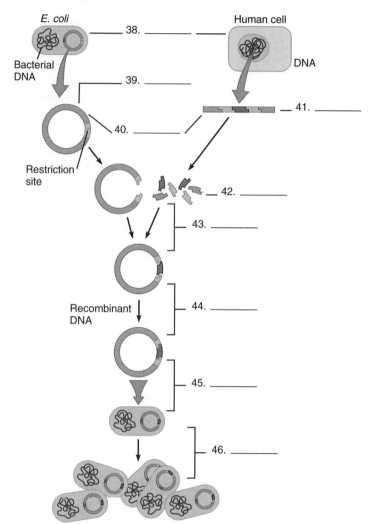

Figure 20.1

Chapter 20 DNA Technology and Genetic Engineering

20.3 Genetic engineering creates transgenic organisms

20.4 Gene therapy: The hope of the future

Crossword Puzzle

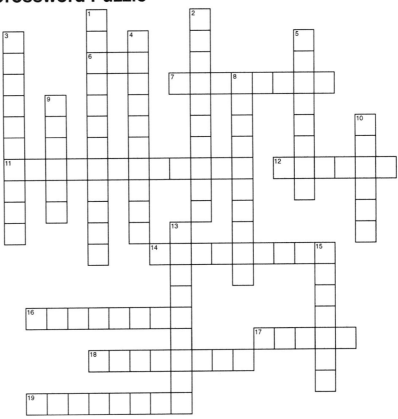

Across

6. Microinjections of DNA into a cell is successful less than ____% of the time.
7. Human insulin is currently produced in genetically engineered _____.
11. Cancer research includes the possibility of adding genes for _____ to a patient's cancer cells.
12. Gene therapy is one of the potiential benefits of the Human _____ Project.
14 Transgenic animals are _____ to produce.
16. When transgenic plants incorporate human genes they produce human _____.
17. Gene therapy for cystic fibrosis may someday involve the use of _____ spray.
18. The process of producing pharmaceuticals in farm animals is called gene _____.
19. Transgenic bacteria aid the production of _____ when they produce foreign antigens.

Down

1. The preferred vector for creating transgenic animals.
2. A genetically engineered vaccine against _____ is currently on the market.
3. The creation of transgenic animals may begin with insertion of foreign DNA into _____ eggs.
4. Viruses used as vectors to create transgenic animals must first be genetically _____.
5. Concerns exist that the use of genetically engineered food crops may lead to some crop _____.
8. _____ organisms carry one or more foreign genes.
9. Something that transports genes into a foreign cell.
10. Gene therapy has successfully treated severe combined _____ deficiency disease.
13. Fatty sphere, which may be able to carry recombinant plasmids into animal cells.
15. Gene _____ involves repairing or replacing damaged human genes.

Chapter Test

Multiple Choice

1. Which of the following functions as a primer in DNA sequencing?
   a. DNA polymerase
   b. nucleotides labeled with a fluorescent marker
   c. short, single-stranded pieces of DNA that bind to the end of a double-stranded piece of DNA
   d. short, double-stranded pieces of DNA that bind to the end of a single-stranded piece of DNA

2. Gel electrophoresis uses _____ to separate DNA pieces according to their _____.
   a. enzymes, bond structure
   b. primers, base sequence
   c. an electrical field, size
   d. a fluoroscope, altered DNA nucleotides

3. All of the following are associated with DNA sequencing except:
   a. DNA nucleotides labeled with a fluorescent marker.
   b. double-stranded DNA primers.
   c. gel electrophoresis.
   d. plasmids incubated with DNA fragments.

4. Recombinant DNA:
   a. has existed for as long as organisms have carried out sexual reproduction.
   b. results from the selective breeding of plants and animals.
   c. produced from the intentional cutting and splicing of DNA molecules has existed for only a few decades.
   d. all of the above.

5. Plasmids occur naturally in:
   a. animal cells.
   b. bacteria.
   c. viruses.
   d. plant cells.

6. Restriction enzymes are most like:
   a. scissors.
   b. tape.
   c. a blueprint.
   d. a copy machine.

7. Cut fragments of DNA can be bound together by:
   a. DNA polymerase.
   b. DNA ligase.
   c. plasmids.
   d. restriction enzymes.

8. Which of the following is *not* required to create a recombinant plasmid?
   a. DNA polymerase
   b. DNA ligase
   c. foreign DNA
   d. restriction enzymes

9. Place the following steps in the order they would occur during the creation of recombinant plasmids.
   I. addition of DNA ligase
   II. isolate DNA from a foreign source
   III. mix foreign DNA fragments with cut plasmids
   IV. introduce recombinant plasmid into bacteria
   V. cut foreign DNA and plasmids with restriction enzymes
   a. I, II, III, IV, V
   b. IV, I, II, V, III
   c. II, V, III, I, IV
   d. II, III, V, I, IV

10. When recombinant plasmids are created to clone human genes:
    a. retroviruses are required to complete the cloning process.
    b. bacteria containing the desired gene must be identified and isolated prior to cloning.
    c. the plasmids are able to reproduce and clone the genes independently, without bacteria.
    d. the gene of interest must first be microinjected into the bacteria.

11. The polymerase chain reaction makes it possible to:
    a. express all the genes in a particular DNA sequence.
    b. amplify a small amount of DNA.
    c. create a large quantity of DNA polymerase for use in genetic engineering.
    d. create transgenic organisms.

12. During PCR, DNA strands are unwound during _____, and complementary strands are formed during _____.
    a. heating, heating
    b. heating, cooling
    c. cooling, heating
    d. cooling, cooling

13. Transgenic organisms are organisms that:
    a. are produced by selective breeding.
    b. carry recombinant plasmids.
    c. are resistant to diseases.
    d. carry one or more foreign genes.

14. Transgenic bacteria are currently used to produce all of the following except:
    a. beta-carotene.
    b. insulin.
    c. human growth hormone.
    d. erythropoietin.

15. Bacteria that produce vaccines have been engineered to express the gene for:
    a. specific antibodies.
    b. human proteins affected by the disease.
    c. the immune cells normally involved in fighting the disease.
    d. the antigen of the disease-causing organism.

16. Using genetically engineered bacteria to produce vaccines is difficult because:
    a. a specific gene must be located and transferred to the bacteria.
    b. the process is time-consuming and expensive.
    c. microorganisms evolve too rapidly to guarantee that a vaccine will always be effective.
    d. all of the above.

17. Transgenic plants:
    a. can be engineered to express bacterial genes, but not human genes.
    b. have been created, but useful applications have not yet been identified.
    c. do not pose any safety or environmental concerns.
    d. have been engineered for herbicide resistance and vitamin delivery.

18. New cancer research involves introduction, by way of a viral vector, of the mda-7 gene into malignant tumors. As a result, malignant tumor cells:
    a. begin to secrete cell-surface proteins that help the immune system to recognize the malignant cells.
    b. are unable to stimulate the growth of the new blood vessels required to nourish the tumor.
    c. lose the ability to metastasize.
    d. die as a result of apoptosis.

19. Transgenic animals are most often created when:
    a. foreign DNA is microinjected into a fertilized egg.
    b. animal cells readily take up recombinant plasmids.
    c. animals are infected by engineered bacteria.
    d. a single defective gene is replaced by a normal gene.

20. Genetic engineering:
    a. is widely accepted as a valuable tool.
    b. has the potential to provide cures for many human health concerns.
    c. raises many ethical problems.
    d. all of the above.

21. The ability to repair or replace damaged human genes is an example of:
    a. gene therapy.
    b. recombinant DNA technology.
    c. genetic engineering.
    d. all of the above.

22. Which of the following is a false statement about gene therapy?
    a. Gene therapy has the potential to help in the treatment of many diseases.
    b. One of the challenges of gene therapy is finding a way to deliver new DNA to all necessary body cells.
    c. Diseases that are successfully treated with gene therapy cannot be inherited by an affected individual's offspring.
    d. Retroviruses are potential vectors for use in gene therapy.

23. The first disease to be treated successfully with gene therapy was:
    a. cystic fibrosis.
    b. SCID.
    c. cancer.
    d. diabetes.

24. Retroviruses have the potential to be effective vectors in delivering DNA to human cells because:
    a. they do not normally infect human cells.
    b. they never mutate.
    c. they are easily microinjected.
    d. they can splice their genetic information into the DNA of human cells.

25. While retroviruses are capable of introducing foreign DNA into a human genome:
    a. this has never been done successfully.
    b. they may disrupt important regulatory genes when they insert foreign DNA at random locations in a chromosome.
    c. their use is limited because they only insert foreign DNA during G1 of the host cell cycle.
    d. the difficulty of creating a recombinant virus makes their common use as a vector unlikely.

Key Concept Review Questions

Each of the Key Concepts listed at the beginning of the study guide has been rewritten as a question below. After successfully completing the study guide exercises and the Chapter Test, you should be able to answer each of these questions. Refer to the Key Concept List at the beginning of this chapter to check your answers.

1. What is biotechnology?
2. What does DNA sequencing allow scientists to determine?
3. What structures, enzymes, and processes are required for gene sequencing?
4. What is recombinant DNA? What three processes are involved in recombinant DNA technology?
5. List four things needed to produce recombinant bacteria.
6. What effect do restriction enzymes have on DNA?
7. What is the action of DNA ligase?
8. What are plasmids and where are they found in nature?
9. What can plasmids carry into bacteria? What must bacteria contain to manufacture a foreign protein?
10. What is PCR?
11. Briefly describe the process of PCR.
12. What are transgenic organisms?
13. What are transgenic bacteria and what can they produce?
14. What are transgenic plants? List some examples of the benefits of transgenic plants.
15. What are some of the concerns associated with the use of transgenic plants?
16. Why does research on transgenic animals continue even though they are difficult to create?
17. What is one technique used to create transgenic animals?

18. What is gene therapy?

19. How are retroviruses used in gene therapy?

20. What disease has been successfully treated with gene therapy? List two diseases that are currently being investigated for treatment with gene therapy.

21

Development and Aging

Chapter Summary and Key Concepts

After reading and studying this chapter you should know the following:

Sections 21.1, 21.2

1. Sperm travel through the vagina, cervix, uterus, and oviduct in an attempt to fertilize the egg.

2. The egg is released as an immature secondary oocyte and completes meiosis II only after fertilization occurs.

3. Fertilization occurs in the upper third of the oviduct when the nuclei of the sperm and ovum fuse to form a zygote.

4. Twins may be fraternal, identical, or conjoined.

5. The four processes of development are: a series of cell divisions called *cleavage,* a process of physical change called *morphogenesis,* the development of specialized form and function in cells called *differentiation,* and an increase in size called *growth*.

Sections 21.3, 21.4

6. Prenatal development is divided into three stages: pre-embryonic, embryonic, and fetal development.

7. In pre-embryonic development, the zygote becomes a morula and then a blastocyst that implants in the endometrium.

8. During embryonic development the embryo consist of three germ layers called the endoderm, the mesoderm, and the ectoderm.

9. Four extraembryonic membranes form during embryonic development to support the embryo: the amnion, the allantois, the yolk sac, and the chorion.

10. The placenta consists of a maternal portion, the endometrium, and an embryonic portion, the chorion and chorionic villi.

11. The umbilical cord connects the placenta to the embryo's circulation.

12. By the end of embryonic development the embryo is distinctly human in appearance.

Sections 21.5, 21.6

13. During fetal development, growth is rapid, the fetus begins to move, surfactant is produced, and life outside the womb becomes possible.

14. Birth involves three stages of labor and delivery: dilation, expulsion, and afterbirth.

15. At birth, changes in the fetal circulation sustain breathing by the newborn.

16. Lactation delivers first colostrum, and then milk, to the newborn.

Sections 21.7, 21.8, 21.9

17. The stages of human development after birth are the neonatal period, infancy, childhood, adolescence, and adulthood.

18. Aging is a process of deterioration that occurs over time.

19. Aging is marked by changes in major body systems.

20. Death is the irreversible cessation of processes required to sustain life.

Exercises

Complete the exercises for each section after you have read and studied the section. If you cannot answer some questions, or answer them incorrectly, return to the chapter and review this information. You may find it helpful to work on only one section at a time. When you have completed all sections, take the Chapter Test as an indicator of your mastery of this topic.

21.1 Fertilization occurs when sperm and egg unite

21.2 Development: Cleavage, morphogenesis, differentiation, and growth

Labeling and Short Answer

Use the terms below to label Figure 21.1 of a sperm and a secondary oocyte, and then answer the questions.

head	nucleus	corona radiata
cytoplasm	midpiece	first polar body
acrosome	plasma membrane	zona pellucida
tail		

1. _____
2. _____
3. _____
4. _____
5. _____
6. _____
7. _____
8. _____
9. _____
10. _____

Figure 21.1

11. What part of a sperm contains enzymes that aid the sperm's entry into an egg?

12. Where does fertilization usually occur?

13. List, in order, the structures in the female through which the sperm must pass before they reach the egg:

14. In the absence of fertilization, how long does the egg survive?

15. What layers of the secondary oocyte must the sperm penetrate to reach the cell membrane?

16. What marks the beginning of fertilization?

17. What events prevent fertilization of the egg by multiple sperm?

18. What event triggers the completion of meiosis II by the secondary oocyte?

19. What is the secondary oocyte called when it completes meiosis II?

20. When is fertilization considered complete?

21. What structures are lost as the ovum becomes a zygote?

Matching

_____ 22. **twins** a. cell divisions without cell growth that occur in the first 4 days after fertilization

_____ 23. **fraternal twins** b. an increase in size that begins with implantation

_____ 24. **identical twins** c. the process of physical changes that occur during development

_____ 25. **conjoined twins** d. twins that arise from the ovulation of more than one oocyte

continued

_____ 26. **cleavage** e. the most common form of multiple births

_____ 27. **morphogenesis** f. the development of specialized forms and functions in individual cells

_____ 28. **differentiation** g. twins that arise when cells begin, but do not complete separation—twins remain attached to each other

_____ 29. **growth** h. twins that arise from a single zygote when cells break apart

21.3 Pre-embryonic development: The first 2 weeks

21.4 Embryonic development: Weeks 3 to 8

Labeling

Use the terms below to label each structure in Figure 21.2.

a. pre-embryo
b. trophoblast
c. endoderm
d. diploid nucleus
e. yolk sac
f. ectoderm
g. inner cell mass
h. morula
i. trophoblast cells
j. zygote
k. embryonic disc
l. blastocyst
m. amniotic cavity
n. embryo
o. hollow cavity
p. endoderm
q. trophoblast cells
r. amniotic cavity
s. embryonic disc
t. mesoderm
u. ectoderm

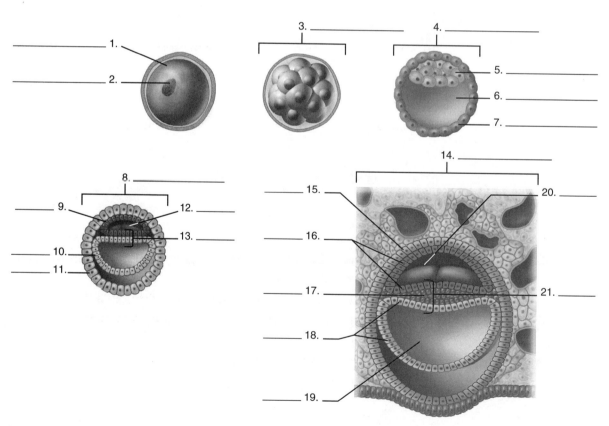

Figure 21.2

Short Answer

Refer to Figure 21.2 as you answer the following questions.

22. What time period after conception is considered to be the pre-embryonic stage?

23. Which structure in Figure 21.2 leaves the oviduct and enters the uterus?

24. Which structure in Figure 21.2 is the first to undergo differentiation and morphogenesis?

25. Which part of the blastocyst will eventually become the embryo?

26. Which structure in Figure 21.2 burrows into the endometrium of the uterus?

27. What is this burrowing process called?

28. What part of the blastocyst secretes enzymes that digest endometrial cells?

29. What marks the end of the pre-embryonic stage?

30. When is the fetus referred to as an embryo?

31. What marks the beginning of the embryonic stage?

32. Which germ layer gives rise to muscle, connective tissue, and bone?

33. Which germ layer gives rise to the lining of the digestive tract and several glands?

34. Which germ layer gives rise to the nervous system and the epidermis of the skin?

Labeling

Use the terms below to label Figure 21.3 below.

uterus   yolk sac   amnion

umbilical cord   amniotic cavity   placenta

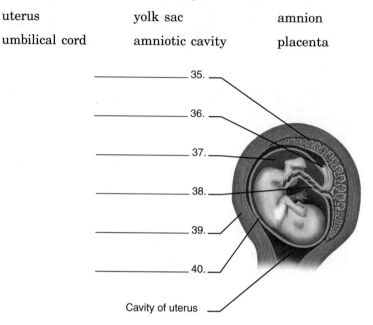

_____ 35.
_____ 36.
_____ 37.
_____ 38.
_____ 39.
_____ 40.

Cavity of uterus

Figure 21.3

Completion

Use the terms below to complete the paragraph, and then repeat this exercise with the terms covered.

trophoblast	amnion	gonads	chorion
endometrium	digestive	yolk sac	umbilical cord
allantois	chorion	embryonic	amnion
endometrium	placenta	villi	umbilical cord

Extra-embryonic membranes and the placenta develop early in the (41)_____ stage. The four extra-embryonic membranes are the (42)_____, (43)_____, (44)_____, and the (45)_____. The (46)_____ is also known as the "bag of waters" and is filled with fluid. The allantois is a temporary membrane that helps form the blood vessels of the (47)_____. The yolk sac becomes part of the fetal (48)_____ tract and provides the germ cells that will move into the (49)_____ and eventually give rise to eggs and sperm. The chorion develops primarily from (50)_____, and will form important structures for gas and nutrient exchange in the (51)_____.

The placenta forms from the (52) _____ of the fetus and the (53) _____ of the mother, and connects to the fetal circulation through the (54) _____. Chorionic cells secrete enzymes that damage the (55) _____ and cause bleeding, providing the medium for gas and nutrient exchange and waste removal for the fetus, while fingerlike projections of the chorion called chorionic (56) _____ contain small capillaries that connect to the umbilical arteries and veins.

21.5 Fetal development: Eight weeks to birth

21.6 Birth and the early postnatal period

21.7 From birth to adulthood

21.8 Aging takes place over time

21.9 Death is the final transition

Short Answer

Write the main events that occur in each month of fetal development below. There are no answers for this exercise in the answer key. Refer to Section 21.5 in the textbook to check your answers.

1. Third month:

2. Fourth month:

3. Fifth month:

4. Sixth month:

5. Seventh through ninth months:

Fill-in-the-Blank

Referenced sections are in parentheses.

6. Labor is triggered by maturation of the fetal _____ _____. (21.6)

7. Contraction of the uterus during labor is a _____ feedback cycle. (21.6)

8. The three phases of labor and delivery are _____, _____, and _____. (21.6)

9. A(n) _____ is a surgical incision that enlarges the vaginal opening for birth. (21.6)

10. In the fetus, blood flows from the umbilical vein to the inferior vena cava through the _____ _____. (21.6)

11. In the fetal circulation, blood flows from the right atrium to the left atrium through the _____ _____. (21.6)

12. After birth, the umbilical blood vessels regress to become _____ tissue. (21.6)

13. The first month of life is referred to as the _____ period. (21.7)

14. The brain grows to 95% of its final size during _____. (21.7)

15. In adolescence, puberty is triggered when the hypothalamus begins secreting _____. (21.7)

16. During the period of _____, a lumbar curve develops and abdominal muscles become stronger. (21.7)

17. A section of disposable DNA found on the end of every DNA strand and believed to play a role in aging is called a _____. (21.8)

18. Menopause occurs when the ovaries lose their responsiveness to _____ and _____. (21.8)

19. Describe in one or two words the changes that occur in each of the following with age: (21.8)

   a. lung tissue   _____
   b. heart walls   _____
   c. bone mass   _____
   d. muscle mass   _____
   e. lens of the eye   _____
   f. viable sperm produced _____
   g. kidney mass   _____
   h. prostate gland   _____

20. Death is defined as the _____ cessation of circulatory and respiratory, or brain, function. (21.9)

Chapter Test

Multiple Choice

1. Fertilization usually occurs in the:
   a. endometrium of the uterus.
   b. ovary.
   c. upper third of the oviduct.
   d. lower third of the oviduct.

2. The protective layer surrounding the secondary oocyte is the:
   a. follicle.
   b. granulosa.
   c. corona radiata.
   d. zona pellucida.

3. Fertilization is considered complete when:
   a. the sperm enters the egg.
   b. the zona pellucida breaks down.
   c. the nuclei of the sperm and ovum unite.
   d. the sperm attaches to the cell membrane of the secondary oocyte.

4. Ovulation of more than one oocyte in a cycle may result in:
   a. fraternal twins.
   b. maternal twins.
   c. conjoined twins.
   d. identical twins.

5. The primary cause of morphogenesis is:
   a. cleavage.
   b. differentiation.
   c. fertilization.
   d. growth.

6. Which of the following is true of the blastocyst?
   a. It develops during the embryonic stage.
   b. It is a ball of 32 identical cells.
   c. It consists of a trophoblast and an inner cell mass.
   d. It develops into a morula.

7. The embryonic disk arises from:
   a. the inner cell mass of the blastocyst.
   b. the inner cell mass of the morula.
   c. the trophoblast.
   d. the ectoderm.

8. Implantation in the endometrium occurs when:
   a. trophoblast cells secrete enzymes that digest endometrial cells.
   b. contact of the blastocyst with the endometrium initiates the growth of new blood vessels.
   c. the morula leaves the oviduct and enters the uterus.
   d. fluid from the amniotic cavity contacts endometrial cells and triggers burrowing of the blastocyst.

9. The nervous system arises from which germ layer?
   a. endoderm
   b. ectoderm
   c. mesoderm
   d. trophoderm

10. The placenta is formed during embryonic development from the:
    a. chorion and endometrium.
    b. amnion and chorion.
    c. amnion and endometrium.
    d. chorion and myometrium.

11. As an endocrine organ, the placenta secretes:
    a. hCG after the corpus luteum degenerates.
    b. progesterone and estrogen to maintain and promote pregnancy when the corpus luteum degenerates.
    c. LH and FSH to stimulate follicle development in the ovaries.
    d. progesterone to maintain the corpus luteum.

12. The neural groove that forms during embryonic development will become:
    a. muscle.
    b. bone.
    c. brain and spinal cord.
    d. heart and blood vessels.

13. It is difficult to distinguish a human embryo from any other vertebrate embryo prior to the _____ week of development.
    a. fourth.
    b. sixth.
    c. eighth.
    d. 12th.

14. Male and female gonads develop in the:
    a. first trimester.
    b. second trimester.
    c. third trimester.
    d. just before birth.

15. A developing human is called a fetus:
    a. after the pre-embryonic stage.
    b. at the beginning of the second trimester.
    c. after the eighth week.
    d. in the third trimester.

16. Contractions of the uterus during labor:
    a. increase the release of oxytocin.
    b. decrease the release of oxytocin.
    c. increase the release of prolactin.
    d. increase the release of ADH.

17. The afterbirth:
    a. is delivered prior to the fetus.
    b. consists of the umbilical cord and the placenta.
    c. is reabsorbed by the uterus.
    d. maintains mild uterine contractions for 24 hours after birth of the fetus to prevent bleeding.

18. Surfactant is important in:
    a. initiating uterine contractions prior to birth.
    b. preventing blood loss when the umbilical cord is severed.
    c. redirecting fetal circulatory pathways in the newborn.
    d. reducing surface tension in the alveoli of the lungs.

19. A shunt that directs fetal circulation from the pulmonary artery to the aorta is the:
    a. ductus arteriosus.
    b. ductus venosus.
    c. foramen ovale.
    d. pulmonary stenosis.

20. Colostrum:
    a. is a pasty substance that covers the newborn's skin.
    b. is a watery milk produced by the breast.
    c. is a fluid lost with the afterbirth.
    d. is an enzyme produced by the myometrium during labor.

21. The period of life from age 2–15 months is termed:
    a. neonatal.
    b. childhood.
    c. infancy.
    d. adolescence.

22. The _____ system lags behind most other systems in its development.
    a. cardiovascular
    b. immune
    c. muscular
    d. nervous

23. All of the following are accepted theories of the cause of aging except:
    a. genetic programming determines the timing of cell death.
    b. environmental toxins have led to a shortened life span.
    c. accumulated cell damage interferes with cell repair.
    d. interdependence of body systems promotes aging when one system begins to fail.

24. The finding that severe caloric restriction in some animals prolongs life supports the theory that aging is due to:
    a. accumulated damage to DNA and errors in DNA replication.
    b. an internal genetic counting mechanism, possibly involving telomeres, that limits the number of cell divisions in each cell.
    c. the decline in one body system that affects other body systems.
    d. poor nutrition.

25. Death may be defined as a cessation of either circulatory or respiratory function or of brain function. In either definition, a key word preceding "cessation" is:
    a. complete.
    b. prolonged.
    c. irreversible.
    d. acute.

Key Concept Review Questions

Each of the Key Concepts listed at the beginning of this chapter has been rewritten as a question below. After successfully completing the study guide exercises and the Chapter Test, you should be able to answer each of these questions. Refer to the Key Concepts list at the beginning of this chapter to check your answers.

1. Trace the path of a sperm from the time it enters the female reproductive tract until it reaches the egg.
2. When does the secondary oocyte complete meiosis II?
3. Where does fertilization usually occur? What is formed by the fusion of the nuclei of the sperm and the ovum?
4. What are the three types of twins?
5. What are the four processes involved in development?
6. What are the three stages of prenatal development?
7. What are the developmental structures that occur in the pre-embryonic stage? Which one becomes implanted in the endometrium?
8. What are the three germ layers that develop during the embryonic stage?
9. Name the four extra-embryonic membranes that develop during the embryonic stage.
10. What structures make up the placenta?
11. What two structures are connected by the umbilical cord?
12. The embryo is distinctly human in appearance by the end of what prenatal stage?
13. What are some of the events that occur during fetal development?
14. What are the three stages of labor and delivery?
15. Is the fetal circulatory system capable of sustaining breathing in the newborn?
16. What substances are provided to the newborn during lactation?
17. What are the five stages of human development that occur after birth?
18. What is aging?
19. Aging is marked by changes in what parts of the body?
20. What is the definition of death?

22

Evolution and the Origins of Life

Chapter Summary and Key Concepts

After reading and studying this chapter you should know the following:

Sections 22.1, 22.2

1. Evolution is the unpredictable and natural process of descent with modification over time.

2. Microevolution occurs as a result of genetic changes.

3. Macroevolution involves large-scale trends in evolution among groups of species.

4. Charles Darwin proposed the theory of descent with modification in the 1800s.

5. Evidence for evolution comes from the fossil record, comparative anatomy and embryology, biochemistry, and biogeography.

6. Fossils are preserved remnants of organisms.

7. The fossil record is incomplete because fossil formation depends on the type of tissue and the environmental conditions.

8. Homologous structures are similar because they evolved from a common ancestor; analogous structures are similar because they evolved for a similar function.

9. Vestigial structures serve little or no function and may be the remains of an ancestral structure.

10. Similarities in vertebrate embryos suggest that all vertebrates evolved from a common ancestor.

11. Similarities in biochemical molecules provide information about how recently species may have diverged.

12. Biogeography is the study of the distribution of plants and animals around the world.

13. Geographic barriers promote evolution.

14. In the process of natural selection, organisms with traits that make them more fit for the environment are able to survive and reproduce.

15. Evolution is the result of the presence of mutations and the process of natural selection.

16. Genetic drift and gene flow can change populations by changing allele frequencies.

Sections 22.3, 22.4, 22.5

17. Life on earth began with the formation of organic molecules that became enclosed in lipid-protein membranes.

18. The appearance of photosynthetic organisms allowed oxygen to accumulate in the atmosphere, making possible the eventual appearance of animals.

Section 22.6

19. Humans are classified as *Homo sapiens*.

20. The evolution of *Homo sapiens* required ancestral species to walk upright and develop a larger brain.

Exercises

Complete the exercises for each section after you have read and studied the section. If you cannot answer some questions, or answer them incorrectly, return to the chapter and review this information. You may find it helpful to work on only one section at a time. When you have completed all sections, take the Chapter Test as an indicator of your mastery of this topic.

22.1 Evidence for evolution comes from many sources

22.2 Natural selection contributes to evolution

continued

Matching

____ 1. **evolution** a. body structures that are similar because they share a common function, not a common origin

____ 2. **microevolution** b. a map that illustrates evolutionary change and relationships among species

____ 3. **macroevolution** c. random changes in allele frequency because of chance effects

____ 4. **fossils** d. the process whereby individuals with traits that make them more fit are more likely to survive and reproduce

____ 5. **homologous structures** e. a group of individuals of the same species that occupy the same geographical area

____ 6. **analogous structures** f. the complete disappearance of a species

____ 7. **vestigial structures** g. an unpredictable and natural process of descent over time with genetic modification

____ 8. **continental drift** h. the movement of alleles into and out of a population due to migration

____ 9. **natural selection** i. development of many new species from a single ancestor in a short period of time

____ 10. **population** j. large-scale evolutionary trends or changes that apply to whole groups of species

____ 11. **genetic drift** k. preserved remnants of organisms

____ 12. **gene flow** l. the movement of plates to which the continents are attached

____ 13. **extinction** m. body structures that are similar because they share a common ancestral origin

____ 14. **evolutionary tree** n. evolution that occurs as a result of genetic changes that give rise to a new species

____ 15. **adaptive radiation** o. body structures that currently have little or no function

Fill-in-the-Blank

Referenced sections are in parentheses.

16. Charles Darwin's phrase for evolution was _____ _____ _____. (Intro)

17. Hard tissues are preserved as fossils only when they are quickly covered with _____ or _____ _____. (22.1)

18. Sedimentary deposits are laid layer upon layer in a process called _____. (22.1)

19. When fossils exist in sedimentary rock, the oldest fossils will be located in the _____ layers of rock. (22.1)

20. Radiometric dating in fossils older than 50,000 years requires the use of isotopes of _____. (22.1)

21. In comparative embryology, the gill arches of fish embryos are homologous to the _____ _____ of human embryos. (22.1)

22. Comparative biochemistry suggest a close relationship between species when they have similar or identical _____ _____. (22.1)

23. One species becomes two different species when evolutionary paths _____. (22.1)

24. _____ is the study of the distribution of plants and animals around the world. (22.1)

25. Two hundred million years ago all continents were joined in one land mass called _____. (22.1)

26. Adaptive radiation often occurs after a _____ _____. (22.2)

27. The _____ _____ alters the gene pool of a population when a major catastrophe wipes out most of the population. (22.2)

Short Answer

Refer to Figure 22.1 below to answer the questions.

28. In figure (a):

    a. The human arm and the bird's wing are _____ structures.

    b. Why is there similarity in the structure of the arm and the wing?

    c. What might explain the differences in these structures?

29. In figure (b):

    a. The bird wing and the insect wing are examples of _____ structures.

    b. Did these two wings evolve from a common structure in a common ancestor?

    c. What might explain the similarities between the two wings?

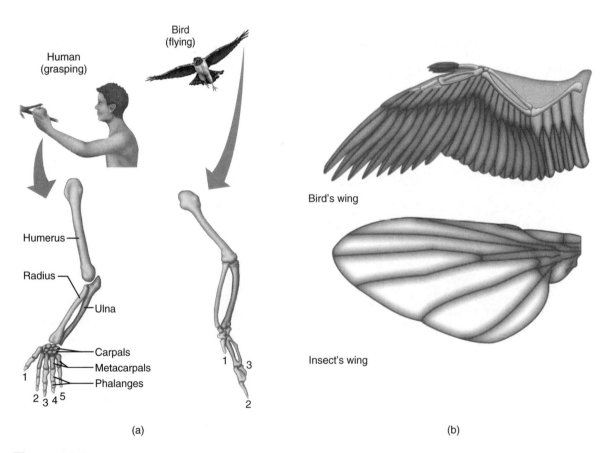

(a) (b)

Figure 22.1

22.3 The young earth was too hot for life

22.4 The first cells were able to live without oxygen

22.5 Photosynthetic organisms altered the course of evolution

Ordering

Label each of the following events with the letters a–k to indicate the sequence in which they occurred as life developed on Earth.

_____ 1. Heat vaporized all water, preventing the formation of oceans.

_____ 2. Oxygen, a by-product of photosynthesis, begins to accumulate in the atmosphere.

_____ 3. Self-replicating RNA molecules form on clay templates along the ocean edge.

_____ 4. Multicellular organisms appear.

_____ 5. Mutation allows some cells to develop the capacity for photosynthesis.

_____ 6. The Earth formed about 4.6 billion years ago with a liquid core of metal and a thin outer crust.

_____ 7. Simple organic molecules of amino acids, sugars, and fatty acids form from molecules in the primitive atmosphere, using heat, radiation, and electrical discharges as energy sources.

_____ 8. New cells evolve with the capacity to use oxygen for aerobic metabolism.

_____ 9. The Earth cools enough for oceans to form.

_____ 10. Simple organic compounds dissolve in the sea.

_____ 11. RNA and small organic molecules become enclosed within a lipid-protein membrane.

Section 22.6 Modern humans came from Africa

Completion

Fill in the table below by writing the scientific name of each taxonomic category, as it applies to humans. Write one or more characteristics that qualify humans for each category.

Taxonomic Category	Scientific Name	Human Characteristics
Kingdom	1.	2.
Phylum	3.	4.
Class	5.	6.
Order	7.	8.
Family	9.	10.
Genus	11.	12.
Species	13.	14.

Labeling

Label Figure 22.2 below, placing each group in the proper evolutionary relationship.

a. *Australopithecus afarensis*
b. *Homo ergaster*
c. *Homo sapiens*
d. Neanderthal
e. *Homo erectus*
f. *Homo heidelbergensis*
g. *Australopithecus anamensis*
h. *Australopithecus africanus*
i. *Homo habilis*

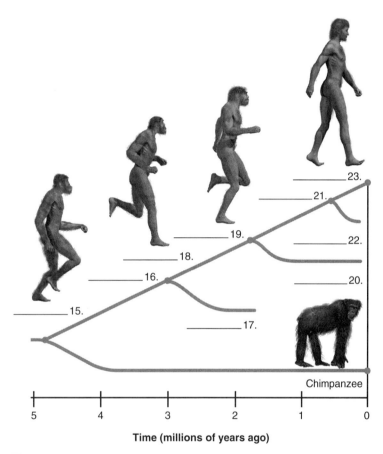

Figure 22.2

Short Answer

24. According to the fossil record, how long ago did humans and chimpanzees share a common ancestor?

25. What was the first group in the lineage that eventually led to humans?

26. What is the defining feature of hominids?

27. According to fossil evidence, when did *Homo sapiens* first evolve?

28. Which species was the first to make tools?

29. Which species arose from *Homo habilis* and migrated out of Africa?

30. A current hypothesis is that *Homo sapiens* evolved from a single colony of what group?

31. *Homo sapiens* have been the sole surviving human species for how long?

32. What were the final three characteristics to develop in the evolution of modern humans?

Chapter Test

Multiple Choice

1. Evolution that occurs as a result of genetic changes that gives rise to a new species is referred to as:
   a. microevolution.
   b. macroevolution.
   c. genetic evolution.
   d. allelic evolution.

2. Which of the following is most likely to form a fossil?
   a. a soft tissue organism buried by volcanic ash
   b. an aquatic organism
   c. a vertebrate organism covered by sediment
   d. an insect ingested by a reptile

3. Stratification contributes to the process of:
   a. microevolution.
   b. fossilization.
   c. embryonic development.
   d. natural selection.

4. Analogous structures:
   a. arise in organisms that have recently evolved from a common ancestor.
   b. have similar structures, but different functions in organisms of different species.
   c. evolve in organisms that inhabit a similar environment.
   d. have no current use in an organism, but had a purpose in an evolutionary ancestor.

5. The half-life of the radioactive isotope of C^{14} is:
   a. 2000 years.
   b. 5700 years.
   c. 1.5 million years.
   d. 1.3 billion years.

6. The coccyx, or tailbone, of a human is a(n) _____ structure.
   a. homologous
   b. vestigial
   c. analogous
   d. divergent

7. The presence of identical or nearly identical biochemical molecules in two species indicates:
   a. recent divergence.
   b. the presence of analogous structures.
   c. that they developed in the same geographical region.
   d. an increased rate of mutation.

8. The presence of homologous structures in a different species suggests that:
   a. both species evolved from a common ancestor.
   b. one species will be more primitive and one will species will be more complex.
   c. both species evolved in a similar environment.
   d. the homologous structures are used for different functions in each species.

9. The effect of the continental drift has been to:
   a. increase the rate of mutation.
   b. decrease the rate of mutation.
   c. isolate related groups of organisms from each other.
   d. maintain Pangaea.

10. A population is a group of individuals:
    a. who share the same alleles.
    b. who occupy the same area.
    c. of the same species who occupy the same area.
    d. of the same species, anywhere on Earth.

11. Random changes in the allele frequency because of chance effects is called:
    a. gene flow.
    b. natural selection.
    c. genetic drift.
    d. genetic change.

12. Changes in the gene pool of a population due to the founder effect occurs when:
    a. new individuals enter or leave a population.
    b. a major catastrophe alters the allele frequency.
    c. a change in the environment changes the definition of "fitness" for a population.
    d. a few individuals leave a population and become the basis of a new population.

13. The largest mass extinction was the:
    a. Triassic.
    b. Jurassic.
    c. Cretaceous.
    d. Diluvian.

14. Short bursts of evolutionary activity are called:
    a. divergence.
    b. adaptive radiation.
    c. natural selection.
    d. selective radiation.

15. The Earth formed about:
    a. 6 billion years ago.
    b. 4.6 billion years ago.
    c. 4 million years ago.
    d. 2 million year ago.

16. Which of the following was not found in the primitive atmosphere of the earth?
    a. CO_2
    b. H_2O
    c. O_2
    d. N_2

17. Before life evolved, simple organic molecules formed from:
    a. the oceans.
    b. atmospheric gases.
    c. the rain.
    d. DNA and RNA.

18. The first self-replicating molecule was probably:
    a. DNA.
    b. RNA.
    c. lipids.
    d. proteins.

19. The evolution of photosynthetic organisms resulted in:
    a. increased variety of proteins.
    b. increased oxygen in the atmosphere.
    c. evolution of DNA.
    d. increased carbon dioxide in the atmosphere.

20. Organisms capable of aerobic metabolism evolved:
    a. prior to photosynthetic organisms.
    b. after photosynthetic organisms.
    c. at the same time as photosynthetic organisms.
    d. as soon as Earth cooled.

21. Humans and chimpanzees diverged from a common ancestor about:
    a. 2 billion years ago.
    b. 5 million years ago.
    c. 50,000 years ago.
    d. 10,000 year ago.

22. All of the following apply to *Australopithecus afarensis* except:
    a. their diet consisted of vegetables.
    b. they are classified as hominid.
    c. they displayed sexual dimorphism.
    d. they gave rise to *Homo erectus*.

23. Humans qualify for the phylum Chordata because they have:
    a. upright posture.
    b. spoken language.
    c. a nerve cord.
    d. mammary glands.

24. Evidence for chimpanzees as our closest relatives comes from:
    a. analogous structures.
    b. the absence of a tail in chimpanzees.
    c. similar thumb structure.
    d. similarities in human and chimpanzee DNA.

25. Hominids are characterized by:
    a. complex language skills.
    b. opposable thumbs.
    c. upright posture.
    d. mammary glands.

Key Concept Review Questions

Each of the Key Concepts listed at the beginning of this chapter has been rewritten as a question below. After successfully completing the study guide exercises and the Chapter Test, you should be able to answer each of these questions. Refer to the Key Concepts list at the beginning of this chapter to check your answers.

1. What is evolution?
2. What causes microevolution to occur?
3. What characterizes macroevolution?
4. Who proposed the theory of descent with modification?
5. What are four areas that provide evidence for evolution?
6. What are fossils?
7. Why is the fossil record incomplete?
8. What are homologous structures? What are analogous structures?
9. What are vestigial structures?
10. What characteristic of vertebrate embryos supports the theory of evolution?
11. What information is gained by analyzing biochemical molecules?
12. What is biogeography?
13. What effect do geographic barriers have on evolution?
14. What is the process of natural selection?
15. What two things contribute most to the process of evolution?
16. How can genetic drift and gene flow change populations?
17. How did life on earth probably first begin?
18. What effect did photosynthetic organisms have on the atmosphere of the primitive earth?
19. What is the genus and species for humans?
20. What two characteristics were required in ancestral organisms for the development of humans?

23

Ecosystems and Populations

Chapter Summary and Key Concepts

After reading and studying this chapter you should know the following:

Sections 23.1, 23.2, 23.3

1. Ecology is the study of the relationships between organisms and their physical environment.//
2. Organisms and environments interact as populations, communities, ecosystems, and the biosphere.
3. Characteristics of populations include habitat, range, size, actual growth rate, and potential for growth.
4. The biotic potential of a population is the maximum rate of growth.
5. Biotic potential is limited by environmental resistance.
6. The population that an ecosystem can support indefinitely is called the carrying capacity.
7. An organism's niche is its role in the community.
8. Competition may result when niches for different species overlap.
9. Succession is the natural sequence of change in a community.
10. Succession is determined by population growth rates, niches occupied by various species, and competition.
11. Succession ends with the establishment of a climax community.
12. An ecosystem consists of living components called the biomass, and non-living components made up of chemical elements and energy.

Sections 23.4, 23.5

13. Energy flow through an ecosystem is governed by the Laws of Thermodynamics.
14. The First Law of Thermodynamics states that energy cannot be created or destroyed.

15. The Second Law of Thermodynamics states that whenever energy changes form or is transferred, some of the energy will be transformed into nonuseful forms.

16. Producers are organisms capable of photosynthesis.

17. Consumers are organisms that must consume food for energy.

18. In an ecosystem, energy flows from the sun to producers, and then to consumers.

19. In a biogeochemical cycle, chemicals cycle between the biomass, exchange pool, and reservoir pool.

20. Important biogeochemical cycles include the cycles of water, carbon, nitrogen, and phosphorus.

Exercises

Complete the exercises for each section after you have read and studied the section. If you cannot answer some questions, or answer them incorrectly, return to the chapter and review this information. You may find it helpful to work on only one section at a time. When you have completed all sections, take the Chapter Test as an indicator of your mastery of this topic.

23.1 Ecosystems: Communities interact with their environment

23.2 Populations: The dynamics of one species in an ecosystem

23.3 Communities: Different species living together

Matching

____ 1. **ecology**   a. the maximum rate of growth for a population

____ 2. **ecosystem**   b. a stable community achieved by succession

____ 3. **population**   c. all the ecosystems of the earth

____ 4. **community**   d. the living components of an ecosystem

____ 5. **biosphere**   e. a natural sequence of change in the organisms that dominate a community

____ 6. **habitat**   f. a group of individuals of the same species that occupy the same geographic area

____ 7. **geographic range** g. an organism's role in the community

____ 8. **biotic potential** h. the study of the relationships between organisms and their physical environment

_____ 9. **carrying capacity**   i. the populations of all species that occupy the same geographic area

_____ 10. **environmental resistance**   j. the area over which a species may be found

_____ 11. **niche**   k. a community of organisms and the physical environment in which they live

_____ 12. **succession**   l. the type of location where a species chooses to live

_____ 13. **climax community**   m. the population that an ecosystem can support indefinitely

_____ 14. **biomass**   n. environmental factors that limit a species' ability to realize its biotic potential

Fill-in-the-Blank

Referenced sections are in parentheses.

15. A species' _____ is determined by its tolerance for environmental conditions. (23.2)

16. Three things that limit an organism's range are _____, _____, and _____. (23.2)

17. A population stabilizes when it reaches a balance between _____ _____ and _____ _____. (23.2)

18. Factors that kill organisms or prevent them from reproducing contribute to _____ _____. (23.2)

19. Every growing population will eventually reach a point where environmental resistance begins to _____. (23.2)

20. _____ _____ occurs when one species out-competes another species in a given location. (23.3)

21. _____ _____ are very efficient in utilitizing energy and nutrients, but are also very vulnerable due to their _____. (23.3)

22. The non-living components of an ecosystem consist of the _____ _____ and _____. (23.3)

Short Answer

Refer to Figure 23.1 to answer the questions.

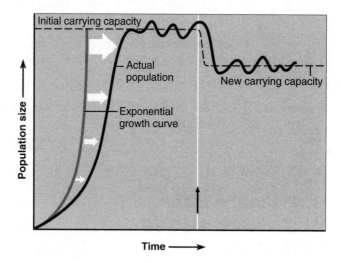

Figure 23.1

23. What kind of growth is reflected by an exponential growth curve?

24. What is the cause of the actual population differing from the exponential growth curve?

25. What causes the carrying capacity of a population to change?

26. What four factors influence the biotic potential of a species?

23.4 Energy flows through ecosystems

23.5 Chemical cycles recycle molecules in ecosystems

Matching

___ 1. **producers** a. indicates total amount of energy at each level of an ecosystem, or total biomass stored at each level

___ 2. **consumers** b. a step in the biogeochemical cycle of nitrogen in which bacteria convert nitrates into atmospheric nitrogen gas

___ 3. **herbivore** c. the process of converting atmospheric nitrogen gas to ammonia

___ 4. **carnivores** d. carbon-containing compounds obtained from fossils and used as an energy source

___ 5. **omnivores** e. organisms that cannot synthesize their own food

___ 6. **decomposers** f. the interaction of multiple food chains

___ 7. **food web** g. consumers that utilize the energy stored in green plants

___ 8. **ecological pyramid** h. consumers that obtain energy from dead organisms

___ 9. **biogeochemical cycle** i. the recycling of chemicals that compose living organisms between the organisms and the earth

___ 10. **fossil fuels** j. consumers that can derive energy from plants or animals

___ 11. **nitrogen fixation** k. the formation of nitrate

___ 12. **nitrification** l. organisms capable of photosynthesis

___ 13. **denitrification** m. consumers that feed on other animals

Labeling

On the left side of Figure 23.2, label the role each organism plays in the food chain. On the right side of the figure, label the characteristics applicable to each organism. Use the lists provided; each characteristic may be used more than once.

Roles

a. secondary consumer
b. producer
c. tertiary consumer
d. primary consumer

Characteristics

a. referred to as an autotroph
b. release CO_2 into the atmosphere
c. herbivore
d. performs photosynthesis
e. release O_2 into the atmosphere
f. carnivore
g. referred to as a heterotroph
h. least efficient use of energy

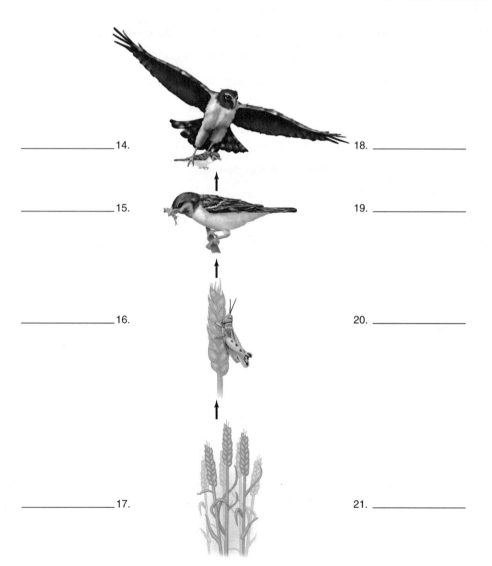

_____ 14. _____ 18.

_____ 15. _____ 19.

_____ 16. _____ 20.

_____ 17. _____ 21.

Figure 23.2

Use the phrases below to label Figure 23.3 of the water cycle.

a. precipitation over land
b. precipitation over ocean
c. ocean to land
d. evaporation from land
e. evaporation from ocean
f. land to ocean

Figure 23.3

Completion

Use the terms below to complete the paragraph, then repeat this exercise with the terms covered.

carbohydrates	atmosphere	atmospheric CO_2 gas
organic	bones	aerobic respiration
shells	CO_2	solar energy
carbohydrates	photosynthesis	aerobic respiration

Carbon forms the backbone of many (28) _____ molecules and the crystallized structure of (29) _____ and (30) _____. Carbon in living organisms is exchanged with (31) _____. The carbon cycle is dependent on (32) _____ by plants and (33) _____ by plants and animals. During photosynthesis, plants use CO_2, (34) _____, and water to form (35) _____, releasing oxygen as a by-product. Plants and animals then utilize (36) _____ with oxygen to break down (37) _____ in plants for energy. This process produces (38) _____, which is released into the (39) _____.

Fill-in-the-Blank

40. The First Law of Thermodynamics states that energy cannot be created or _____. (23.4)

41. Small organelles called _____ are found in plant cells and house the reactions of photosynthesis. (23.4)

42. During the _____ cycle of photosynthesis, plants use CO_2 from the atmosphere to manufacture carbohydrates. (23.4)

43. In a biogeochemical cycle, producers obtain nutrients from the _____ pool, and incorporate them into the _____. (25.5)

44. Nitrogen fixation is carried out by _____. (23.5)

45. All organisms, other than plants, must rely on _____ for usable nitrogen. (23.5)

46. In denitrification, denitrifying bacteria convert some nitrates to _____ _____. (23.5)

47. The phosphorus cycle is called a _____ cycle because phosphorus never enters the atmosphere. (23.5)

Short Answer

48. Why is energy described as flowing through the ecosystem in one direction, rather than being recycled?

49. Why are nutrients described as cycling through the ecosystem, rather than flowing in one direction?

Chapter Test

Multiple Choice

1. The populations of all the species living in a specific geographical area is the:
   a. ecosystem.
   b. community.
   c. biosphere.
   d. composite population.

2. Ecology is best defined as the study of:
   a. the cycling of elements in the environment.
   b. the relationship between organisms and their physical environment.
   c. ecological pyramids.
   d. fossil fuels.

3. The biosphere consists of:
   a. all the living things on the earth.
   b. all the living and nonliving things in a geographical area.
   c. all the ecosystems of the earth.
   d. all the nonliving things on the earth.

4. Which of the following does not limit an organism's geographic range?
   a. competition for resources
   b. reproductive behaviors
   c. extreme or intolerable conditions
   d. physical obstacles and geographic barriers

5. The biotic potential of an organism will usually:
   a. increase without limits.
   b. increase independently of carrying capacity.
   c. not be affected by environmental resistance.
   d. follow an exponential growth curve.

6. One determinant of the biotic potential of a population is:
   a. time required for offspring to reach reproductive maturity.
   b. genetic compatibility.
   c. gene flow and genetic drift.
   d. environmental conditions and geographic barriers.

7. Environmental resistance is best defined as:
   a. the ability of an organism to adapt to a wide range of environments.
   b. the ability of an organism to overcome environmental obstacles to reproduction.
   c. factors that limit a species' ability to realize its biotic potential.
   d. the number of populations competing for the same resources.

8. The population that an environment can support indefinitely is called the:
   a. climax community.
   b. carrying capacity.
   c. maximum population.
   d. environmental potential.

9. The carrying capacity of an ecosystem is achieved when:
   a. a balance occurs between biotic potential and environmental resistance.
   b. the reproductive rate of a population declines.
   c. environmental resistance increases.
   d. competitive exclusion increases.

10. Competitive exclusion could not occur without:
    a. two or more populations competing for the same resources.
    b. two or more organisms of the same species competing for the same resources.
    c. more organisms of reproductive age than the environment is capable of supporting.
    d. a change in the physical environment that limits the availability of resources.

11. An organism's functional relationship with all living and nonliving resources in its geographical area is its:
    a. habitat.
    b. niche.
    c. community.
    d. biome.

12. In photosynthesis, the products of the Calvin cycle include:
    a. O_2 and CO_2.
    b. O_2 and sugar.
    c. sugar and ATP.
    d. sugar and ADP.

13. Chlorophyll is:
    a. a plant cell that carries out photosynthesis.
    b. a species of algae that functions as a producer in an aquatic ecosystem.
    c. a light-absorbing pigment.
    d. a storage organelle for excess carbohydrates in plant cells.

14. Climax communities are:
    a. the most efficient in energy and nutrient utilization.
    b. occupied by one species.
    c. not involved in competition.
    d. only found in certain locations.

15. The total living component of an ecosystem is called the:
    a. climax.
    b. food web.
    c. community.
    d. biomass.

16. A constant supply of energy is essential for continued life because the transfer of energy from one organism to another organism is an energy-_____ process.
    a. absorbing
    b. producing
    c. releasing
    d. balanced

17. The chemical matter of the earth is _____ the earth and the biomass.
    a. transferred to
    b. divided between
    c. produced by
    d. recycled between

18. The Second Law of Thermodynamics states that:
    a. energy cannot be created or destroyed.
    b. energy can be changed from one form to another.
    c. energy can be used, or stored in a usable form.
    d. energy transfer is inefficient.

19. Humans can function as all of the following except:
    a. secondary consumers.
    b. tertiary consumers.
    c. omnivores.
    d. autotrophs.

20. The population of consumers at any level of a food chain depends most directly on:
    a. the amount of energy available from the sun.
    b. the biogeochemical cycles of essential chemicals.
    c. the population of consumers directly below it.
    d. the vulnerability of the producers.

21. Molecules and elements in living things cycle between the biomass, a reservoir, and a(n):
    a. producer.
    b. consumer.
    c. exchange pool.
    d. recycling pool.

22. The total amount of energy stored at each level of an ecosystem is depicted in a(n):
    a. food web.
    b. food chain.
    c. ecological pyramid.
    d. exchange cycle.

23. Which of the following would not affect the carbon cycle?
    a. photosynthesis by plants
    b. aerobic respiration by plants and animals
    c. sunlight
    d. atmospheric phosphorus

24. Living things depend on nitrogen fixation:
    a. to detoxify atmospheric nitrogen.
    b. to convert atmospheric nitrogen to a usable form.
    c. to make nitrogen available for aerobic respiration.
    d. to return nitrogen to the atmosphere.

25. The phosphorus cycle is a sedimentary cycle because:
    a. phosphorus cycles through only two pools.
    b. phosphorus is obtained from sediments.
    c. the phosphorus cycle relies on water availability.
    d. phosphorus never enters the atmosphere.

Key Concept Review Questions

Each of the Key Concepts listed at the beginning of the study guide has been rewritten as a question below. After successfully completing the study guide exercises and the Chapter Test, you should be able to answer each of these questions. Refer to the Key Concept List at the beginning of this chapter to check your answers.

1. What is ecology?
2. List four levels of interaction involving organisms with each other or with the environment.
3. List five characteristics of populations.
4. What is the biotic potential of a population?
5. What limits the biotic potential of a population?
6. What is carrying capacity?
7. What is an organism's niche?
8. What happens when the niches for two different species overlap?
9. What is succession?
10. What factors influence succession?
11. What is the culmination of succession?
12. What are the living and non-living components of an ecosystem?
13. What laws govern the flow of energy through an ecosystem?
14. What is the First Law of Thermodynamics?
15. What is the Second Law of Thermodynamics?
16. What are producers?
17. What are consumers?
18. What is the sequence of energy flow in an ecosystem between producers, consumers, and the sun?
19. What happens in a biogeochemical cycle?
20. List four common and important chemicals involved in biogeochemical cycles.

24

Human Impacts, Biodiversity, and Environmental Issues

Chapter Summary and Key Concepts

After reading and studying this chapter you should know the following:

Section 24.1

1. About 5000 years ago the development of agriculture and better living conditions allowed the human population to rise.

2. Rapid growth began in the 1700s with the Industrial Revolution.

3. The human population has currently entered a phase of explosive growth.

4. The fertility rate is the number of children born to each woman; the replacement fertility rate is the average fertility rate required to reach zero population growth.

5. Fertility rates differ in different regions of the world. To reach zero population growth, the fertility rate must fall to the replacement fertility rate.

6. A population can continue to grow after reaching the replacement fertility rate, depending on its age structure.

7. The age structure of a population is linked to economic development.

8. Industrialization raises human carrying capacity and allows population growth.

Sections 24.2, 24.3, 24.4

9. The primary greenhouse gases are carbon dioxide, methane, and the air pollutants CFCs.

10. The greenhouse effect allows sunlight to pass while trapping heat. This is a normal process and helps to maintain the temperature of the earth.

11. Global warming is an increase in average global temperatures and occurs when excess greenhouse gases are produced. Two human activities, deforestation and the burning of fossil fuels, contribute to the potential for global warming.

12. Ozone occurs near the earth's surface as a result of air pollution, and in the stratosphere where it is protective. CFCs deplete the stratospheric ozone layer.

13. Acid rain is produced when air pollutants combine with water in the atmosphere.

14. Smog occurs when air pollutants react with each other, and usually appears after an area has experienced rapid industrial growth.

15. Fresh water comprises less than 1% of the earth's water, and is distributed unequally.

16. Human activities can lead to pollution of surface water, ground water, and ocean water.

17. The expansion of human populations and the search for resources has polluted existing land and led to a reduction in productive land.

Sections 24.5, 24.6, 24.7

18. Humans have high energy demands and inadequate energy resources.

19. Current energy sources are primarily fossil fuels; additional energy sources may include nuclear energy, hydroelectric power, biomass fuel, and solar energy.

20. A sustainable world is one in which humans and the ecosystems are maintained.

Exercises

Complete the exercises for each section after you have read and studied the section. If you cannot answer some questions, or answer them incorrectly, return to the chapter and review this information. You may find it helpful to work on only one section at a time. When you have completed all sections, take the Chapter Test as an indicator of your mastery of this topic.

24.1 Human population growth: A relatively recent issue

Matching

___ 1. **growth rate**
___ 2. **zero population growth**
___ 3. **fertility rate**
___ 4. **replacement fertility rate**
___ 5. **demographic transition**
___ 6. **more industrialized countries (MICs)**
___ 7. **less industrialized countries (LICs)**

a. the point at which births equal deaths
b. the average fertility rate required to achieve zero population growth
c. nations that are just beginning to industrialize
d. the number of births per year minus the number of deaths per year, divided by the population size
e. nations with established industry-based economies
f. the number of children born to each woman during her lifetime
g. a change in a human population over time

Fill-in-the-Blank

Referenced sections are in parentheses.

8. The global human population first began to rise slowly with agricultural development and conditions that led to a(n) _____ in environmental resistance and a(n) _____ in carrying capacity. (24.1)

9. The total human growth rate has been _____ since 1965. (24.1)

10. In order to reach zero population growth, the human population will have to either _____ the birth rate or _____ the death rate. (24.1)

11. The current human fertility rate is about _____ children per woman, while the replacement fertility rate is _____ children per woman. (24.1)

12. The fertility rate is highest in _____ industrialized countries. (24.1)

13. _____ is the study of human populations. (24.1)

14. The demographic transition of a region is related to its _____ development and the _____ well-being of its citizens. (24.1)

15. In the first stage of demographic transition, a society will have a(n) _____ death rate and a(n) _____ birth rate. (24.1)

16. In the final stages of demographic transition, the birth rate _____ the death rate. (24.1)

Short Answer

Refer to Figure 24.1 to answer the questions.

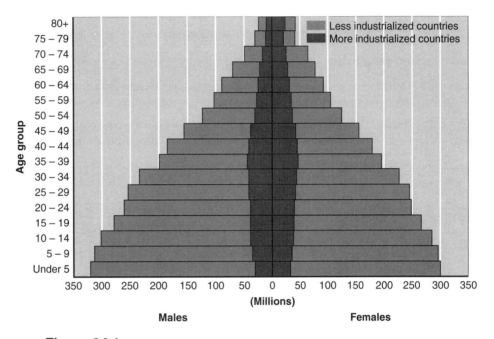

Figure 24.1

17. What is depicted in this figure?

18. Which population has more women of reproductive age?

19. Why will the less industrialized countries have a difficult time controlling population growth even if they have achieved replacement fertility rate?

20. This diagram suggests that industrialization affects what feature of a population?

24.2 Pollutants impair air quality

24.3 Pollution jeopardizes scarce water supplies

24.4 Pollution and overuse damage the land

Matching

____ 1. **pollutants**
____ 2. **greenhouse effect**
____ 3. **global warming**
____ 4. **deforestation**
____ 5. **ozone**
____ 6. **smog**
____ 7. **thermal inversion**
____ 8. **aquifers**
____ 9. **eutrophication**
____ 10. **biological magnification**
____ 11. **desertification**

a. a condition that occurs when a warm upper layer of air traps a cooler air mass containing smog beneath it
b. the accumulation of organic and inorganic nutrients leads to the growth of plant life and death of animal life
c. the transformation of marginal lands into near-desert conditions
d. deep underground reservoirs of fresh water
e. chemicals in the environment that have adverse effects on living organisms
f. a condition where toxic pollutants become more concentrated in the tissue of animals higher in the food chain
g. when sunlight is allowed to pass through the atmosphere but most of the heat is trapped
h. removing trees from large areas of land
i. a raise in the average global temperature
j. O_3 found in two places in the atmosphere
k. a brown or gray layer formed in the atmosphere when air pollutants react with each other

Short Answer

Refer to Figure 24.2 to answer the following questions.

12. Which pathways of the carbon cycle have been affected by human activity in such a way that atmospheric levels of CO_2 are increasing?

13. Describe the activity associated with each pathway and how it is causing CO_2 levels to rise.

14. Does this increase in CO_2 levels increase or decrease the greenhouse effect?

15. What effect can this rising CO_2 level have on the temperature of the earth?

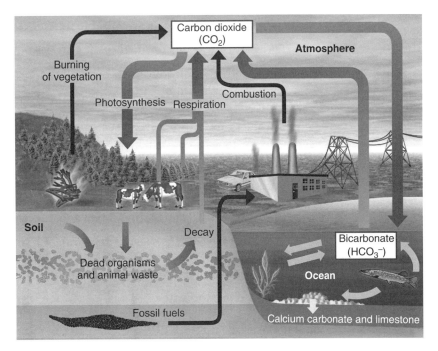

Figure 24.2

Word Choice

Circle the answer that correctly completes the sentence.

16. Ozone in the ____(troposphere/stratosphere)____ helps to protect the earth from UV radiation.

17. Ozone in the ____(troposphere/stratosphere)____ is a pollutant formed when oxygen reacts with automobile exhaust and industrial pollution.

18. CFCs ____(increase/decrease)____ the level of ozone in the stratosphere.

19. CFCs ____(increase/decrease)____ global warming.

20. Acid rain may form when ____(chlorine/sulfur dioxide)____ is released into the air.

21. Nitrogen oxides and hydrocarbons are the primary contributors to ____(global warming/smog formation)____.

22. Freshwater comprises ____(less than/more than)____ 1% of the earth's total amount of water.

continued

23. Water pollutants that are organic nutrients __(increase/decrease)__ the growth rate of bacteria, resulting in a(n) __(increased/decreased)__ availability of oxygen for other aquatic organisms.

24. Eutrophication __(converts freshwater into/protects freshwater from becoming)__ a marsh.

25. Most toxic pollutants __(can/cannot)__ be degraded by biological decomposition.

26. In biological magnification, toxic substances become __(more/less)__ concentrated in the tissues of animals higher up in the food chain.

27. Many of the pollutants that affect surface water __(do/do not)__ pollute groundwater.

28. Desertification __(increases/decreases)__ the amount of available productive land.

24.5 Energy: many options, many choices

24.6 Human impacts are creating a biodiversity crisis

24.7 Toward a sustainable future

Completion

Fill in the table below, listing the disadvantages of utilizing various sources of energy.

Energy Source	Disadvantages
Fossil Fuels	1.
	2.
	3.
	4.
Nuclear Energy	5.
	6.
	7.

Energy Source	Disadvantages
Hydroelectric Power	8.
Biomass Fuels	9.
	10.

Short Answer

11. Solar power is a renewable energy source. What currently prevents solar power from becoming our primary energy source? (24.5)

12. What human activities have contributed to a reduction in biodiversity? (24.6)

13. What are four things we can do to work toward creating a sustainable world? (24.7)

Chapter Test

Multiple Choice

1. The global human population first began to rise about:
   a. 500,000 years ago.
   b. 5,000 years ago.
   c. 500 years ago.
   d. 300 year ago.

2. The rapid rise in human population can be attributed to an increase in all of the following except:
   a. industrialization
   b. availability of antibiotics and vaccines
   c. environmental resistance
   d. geographic range

3. Rapid growth in the human population began in the:
   a. 1200s.
   b. 1500s.
   c. 1700s.
   d. 1900s.

4. Industrialization raises the _____ of a human population.
   a. biotic potential
   b. environmental resistance
   c. reproductive rate
   d. carrying capacity

5. The growth rate of a population is calculated as the number of births per year minus the number of deaths per year, divided by the:
   a. fertility rate.
   b. replacement fertility rate.
   c. current growth rate.
   d. population size.

6. A population can fail to have zero population growth even after reaching the replacement fertility rate due to:
   a. age structure.
   b. infant mortality rate.
   c. number of elderly in the population.
   d. environmental conditions.

7. Which of the following is not considered to be a more industrialized country?
   a. North America
   b. Latin America
   c. Northeastern Asia
   d. Australia

8. The highest growth rates in human populations are seen in:
   a. MICs.
   b. LICs.
   c. there is no distinguishable pattern.
   d. countries with higher economic and educational levels.

9. Which of the following is not a primary component of the air we breathe?
   a. nitrogen
   b. oxygen
   c. carbon dioxide
   d. magnesium

10. The greenhouse effect:
    a. prevents sunlight from passing.
    b. allows heat to escape.
    c. is important in maintaining normal surface temperatures on the earth.
    d. is not involved in global warming.

11. The two main human activities that raise the level of CO_2 in the atmosphere are:
    a. deforestation.
    b. desertification.
    c. burning of fossil fuels.
    d. a and c.
    e. b and c.

12. Ozone in the atmosphere is:
    a. beneficial in the troposphere and harmful in the stratosphere.
    b. beneficial in the stratosphere and harmful in the troposphere.
    c. beneficial in both the troposphere and the stratosphere.
    d. harmful in both the troposphere and the stratosphere.

13. Which of the following does not contribute directly to the formation of acid rain?
    a. burning high-sulfur coal
    b. automobile exhaust
    c. burning oil
    d. deforestation

14. Acid rain is produced when:
    a. holes in the ozone layer affect atmospheric gases.
    b. sulfur dioxide and nitrogen oxides combine with water vapor in the air.
    c. water in the atmosphere ionizes.
    d. industrial plants release acid vapors into the air.

15. What condition can be worsened by a thermal inversion?
    a. ozone depletion
    b. smog
    c. acid rain
    d. global warming

16. Water pollutants that are inorganic nutrients:
    a. cause prolific growth of algae.
    b. cause prolific growth of bacteria.
    c. are degraded by bacteria.
    d. arise from sewage treatment plants.

17. Eutrophication:
    a. refers to the rapid growth of animal life in a shallow body of water.
    b. results from a depletion of organic nutrients.
    c. is accelerated by water pollution.
    d. converts dry land into marsh.

18. In biological magnification, _____ become more concentrated in animals that are _____.
    a. toxic substances, higher up in the food chain
    b. toxic substances, lower in the food chain
    c. nutrients, higher up in the food chain
    d. nutrients, lower in the food chain

19. Heat pollution in water:
    a. increases the amount of oxygen in the water.
    b. decreases the oxygen demand of some aquatic organisms.
    c. may lead to suffocation of some aquatic organisms.
    d. protects freshwater from organic contaminants.

20. It is difficult to determine the full effects of groundwater contamination because:
    a. the effects may be confused with other causes of human disease.
    b. it is difficult to detect the presence of radioactive waste.
    c. groundwater is protected from most contaminants by its location.
    d. many effects of groundwater contamination are never reported.

21. About half of the oil that enters the ocean every year comes from:
    a. oil disposal on land that washes out to sea.
    b. accidents at sea.
    c. natural seepage.
    d. acid rain.

22. Which of the following contributes least to the damage and overuse of land?
    a. reliance on the land as a resource in rural communities
    b. expansion of cities
    c. urban poverty
    d. war

23. Which of the following is a true statement about fossil fuels?
    a. They are one fuel source that does not have an environmental cost.
    b. Use of fossil fuels may reduce the level of greenhouse gases.
    c. Use of fossil fuels may contribute to acid rain.
    d. They are a nonrenewable energy source.

24. Examples of biomass fuels include:
    a. wood.
    b. coal.
    c. oil.
    d. water.

25. Which of the following would not contribute to the establishment of a sustainable world?
    a. recycling
    b. reducing consumption
    c. increasing the fertility rate
    d. reducing poverty

Key Concept Review Questions

Each of the Key Concepts listed at the beginning of the study guide have been rewritten as a question below. After successfully completing the study guide exercises and the Chapter Test, you should be able to answer each of these questions. Refer to the Key Concept List at the beginning of this chapter to check your answers.

1. How long ago did the human population begin to rise? What led to this rise?
2. When did rapid growth in the human population begin? What event did this coincide with?
3. What is the current state of human population growth?
4. How is the fertility rate of a population determined? What is the replacement fertility rate?
5. How do fertility rates compare around the world? What fertility rate coincides with zero population growth?
6. What can cause a population to continue to grow after reaching the replacement fertility rate?
7. What is the correlation between age structure and economic development?
8. What is the effect of industrialization on human carrying capacity and population growth?
9. What are the primary greenhouse gases?
10. What is the greenhouse effect? Is it a normal process? What aspect of the environment do greenhouse gases help to maintain?
11. What is global warming? What may cause global warming? What two human activities contribute to the potential for global warming?
12. Where is ozone found? In which area is ozone beneficial? What chemicals contribute to depletion of the ozone layer?
13. What causes acid rain?
14. What causes smog?
15. How much of the earth's total water supply is freshwater? Is freshwater distributed equally over the earth?
16. What sources of water can be polluted by human activities?

17. What effect have human activities had on existing land and available productive land?

18. How would you characterize the energy demands of human populations?

19. What is currently the primary energy source for human populations? Name four additional energy sources.

20. What is a sustainable world?

Answer Key

Chapter 1

Sections 1.1, 1.2

1. m; **2.** d; **3.** i; **4.** l; **5.** j; **6.** c; **7.** k; **8.** f; **9.** g; **10.** h; **11.** a; **12.** e; **13.** b; **14.** prokaryotes; **15.** eukaryotes; **16.** *Animalia, Plantae, Fungi, Protista, Monera;* **17.** vertebrates; **18.** mammals; **19.** primates; **20.** bipedalism; **21.** opposable; **22.** brain; **23.** language; **24.** living things have a different molecular composition than nonliving things; **25.** populations of living things evolve; **26.** living things are composed of cells; **27.** living things require energy and raw materials; **28.** living things respond to their external environment; **29.** living things maintain homeostasis; **30.** living things grow and reproduce; **31.** h; **32.** c; **33.** a; **34.** f; **35.** d; **36.** b; **37.** g; **38.** e; **39.** i; **40.** while individual organisms may not show every characteristic, populations of similar organisms do; **41.** domain encompasses kingdoms; **42.** bipedalism, opposable thumbs, large brain, complex language ability

Sections 1.3, 1.4, 1.5, 1.6

1. g; **2.** k; **3.** h; **4.** i; **5.** c; **6.** j; **7.** e; **8.** b; **9.** a; **10.** f; **11.** d; **12.** observation, experimentation; **13.** inductive; **14.** testable; **15.** prediction; **16.** deductive; **17.** experimental; **18.** blind; **19.** double-blind; **20.** hypothesis; **21.** proven, supported, disproven; **22.** elimination; **23.** peer-reviewed; **24.** data; **25.** standard error; **26.** anecdotal; **27.** fact; **28.** correlation, causes; **29.** observable, natural; **30.** a; **31.** d; **32.** b; **33.** c; **34.** e

Chapter Test

1. c; **2.** e; **3.** c; **4.** d; **5.** c; **6.** d; **7.** d; **8.** c; **9.** c; **10.** b; **11.** b; **12.** b; **13.** b; **14.** a; **15.** d; **16.** d; **17.** b; **18.** c; **19.** d; **20.** c; **21.** b; **22.** c; **23.** b; **24.** c; **25.** d

Chapter 2

Sections 2.1, 2.2

1. d; **2.** f; **3.** i; **4.** m; **5.** k; **6.** o; **7.** a; **8.** p; **9.** n; **10.** b; **11** g; **12.** j; **13.** h; **14.** r; **15.** q; **16.** c; **17.** l; **18.** e; **19.** eight; **20.** shells; **21.** atomic number; **22.** atomic mass; **23.** protons; **24.** radioisotopes; **25.** potential; **26.** orbital; **27.** electrolytes; **28.** polar; **29.** oxygen, carbon, hydrogen, nitrogen, calcium, phosphorus; **30.** double; **31.** ionic; **32.** hydrogen; **33.** c; **34.** a; **35.** f; **36.** e; **37.** d; **38.** b

Sections 2.3, 2.4

1. b; **2.** e; **3.** d; **4.** a; **5.** c; **6.** b; **7.** f; **8.** c; **9.** e; **10.** a; **11.** d

Chapter 2, continued

Sections 2.5, 2.6, 2.7

1. e; 2. g; 3. i; 4. m; 5. k; 6. a; 7. n; 8. l; 9. b; 10. c; 11. h; 12. o; 13. j; 14. d; 15. f; 16. macromolecules; 17. requires; 18. releases; 19. carbohydrates, lipids, proteins, nucleic acids; 20. glucose; 21. c; 22. g; 23. i; 24. e; 25. a; 26. d; 27. f; 28. h; 29. k; 30. j; 31. b

Sections 2.8, 2.9

1. d; 2. g; 3. e; 4. f; 5. b; 6. h; 7. a; 8. c; 9. product; 10. sugar, phosphate, base; 11. RNA; 12. protein; 13. adenine, thymine, cytosine, guanine; 14. single, ribose; 15. ADP, P, energy; 16. amino acids; 17. covalent; 18. amino; 19. carboxylic acid; 20. R; 21. structure; 22. primary; 23. secondary; 24. beta-pleated sheet; 25. tertiary; 26. disulfide; 27. quaternary; 28. polar; 29. denaturation; 30. temperature; 31. pH; 32. c; 33. f; 34. b; 35. h; 36. d; 37. e; 38. a; 39. g

Chapter Test

1. d; 2. d; 3. b; 4. c; 5. b; 6. d; 7. b; 8. d; 9. a; 10. c; 11. a; 12. a; 13. c; 14. d; 15. b; 16. d; 17. b; 18. a; 19. d; 20. c; 21. d; 22. c; 23. b; 24. a; 25. d

---

Chapter 3

Sections 3.1, 3.2, 3.3, 3.4

1. e; 2. h; 3. m; 4. b; 5. s; 6. q; 7. p; 8. a; 9. i; 10. o; 11. d; 12. f; 13. j; 14. g; 15. k; 16. l; 17. n; 18. c; 19. r; 20. nucleus, cytoplasm, plasma membrane; 21. microvilli; 22. transmission electron; 23. lipid bilayer; 24. fluid mosaic; 25. gradient; 26. equilibrium; 27. osmotic pressure; 28. ADP, P, energy; 29. extracellular; 30. activity; 31. b; 32. g; 33. d; 34. j; 35. e; 36. a; 37. c; 38. f; 39. h; 40. i; 41. b, c, f; 42. b, d, f; 43. b, e, f; 44. b, c, d, f, i; 45. a, e, g; 46. a, h; 47. a, h; 48. small; 49. opposite to; 50. cannot; 51. low, high; 52. closed, open; 53. does not; 54. out of, into; 55. out of

Sections 3.5, 3.6

1. mitochondria; 2. cytoskeleton; 3. Golgi; 4. peroxisomes; 5. ribosomes; 6. lysosomes; 7. rough; 8. flagellum; 9. centrioles; 10. nucleus; 11. vesicle; 12. smooth; 13. cilia; 14. reticulum; 15. ribosome; 16. cytosol; 17. cytoskeleton; 18. centriole; 19. mitochondria; 20. lysosome; 21. nuclear membrane; 22. nucleolus; 23. nuclear pore; 24. nucleus; 25. rough ER; 26. smooth ER; 27. Golgi apparatus; 28. secretory vesicle; 29. peroxisome; 30. plasma membrane

Section 3.7

1. e; 2. i; 3. b; 4. g; 5. h; 6. c; 7. f; 8. a; 9. d; 10. mitochondria; 11. carbohydrates; 12. lipids; 13. proteins; 14. cytoplasm; 15. pyruvate; 16. two; 17. pyruvate; 18. acetyl CoA; 19. acetyl groups; 20. coenzyme A; 21. cyclic; 22. acetyl groups; 23. hydrogen; 24. NADH; 25. FADH2; 26. two; 27. hydrogen; 28. inner; 29. energy; 30. active transport; 31. gradient; 32. inner compartment; 33. ATP synthase; 34. it is converted to lactic acid instead of acetyl CoA; 35. oxidative phosphorylation is the only process that requires oxygen and uses energy associated with electron transport; 36. they will be accepted by oxygen in the inner compartment to form water molecules; 37. fats contain long chains of carbon atoms that are broken down into many acetyl groups that can enter the Krebs cycle

Chapter Test

1. b; **2.** b; **3.** d; **4.** c; **5.** a; **6.** d; **7.** b; **8.** c; **9.** a; **10.** d; **11.** d; **12.** c; **13.** d; **14.** c; **15.** b; **16.** d; **17.** b; **18.** a; **19.** d; **20.** b; **21.** a; **22.** d; **23.** b; **24.** a; **25.** d

Chapter 4

Sections 4.1, 4.2, 4.3

1. f; **2.** h; **3.** l; **4.** p; **5.** k; **6.** o; **7.** c; **8.** m; **9.** i; **10.** a; **11.** q; **12.** g; **13.** d; **14.** b; **15.** e; **16.** j; **17.** n; **18.** a) simple squamous, b) permits exchange of nutrients, wastes, and gases; **19.** a) simple cuboidal; b) secretes and reabsorbs water and small molecules; **20.** a) simple columnar, b) absorbs nutrients and produces mucus; **21.** a) exocrine gland, b) secretes product into a hollow organ or duct; **22.** a) stratified squamous, b) protection; **23.** a) stratified cuboidal, b) secretes water and ions; **24.** a) stratified columnar, b) secretes mucus; **25.** a) endocrine gland, b) secretes hormones into the bloodstream; **26.** connective; **27.** matrix; **28.** fibers; **29.** special; **30.** gel-like; **31.** collagen; **32.** elastic; **33.** reticular; **34.** fibroblast; **35.** loose; **36.** dense; **37.** collagen; **38.** elastic; **39.** reticular; **40.** collagen; **41.** chondroblasts; **42.** solid; **43.** blood; **44.** adipocytes; **45.** fat

Sections 4.4, 4.5

1. c; **2.** e; **3.** f; **4.** b; **5.** g; **6.** d; **7.** a; **8.** a) heart, b) one, c) involuntary; **9.** a) attached to bones, b) multiple, c) voluntary; **10.** a) hollow organs and tubes, b) one, c) involuntary

Section 4.6

1. organ; **2.** organ; **3.** tissue; **4.** membrane; **5.** synovial; **6.** serous; **7.** cutaneous; **8.** mucous; **9.** transverse; **10.** proximal; **11.** pleural, pericardial; **12.** anterior; **13.** diaphragm; **14.** respiratory; **15.** muscular; **16.** circulatory; **17.** reproductive; **18.** urinary; **19.** nervous; **20.** skeletal; **21.** integumentary; **22.** lymphatic; **23.** endocrine; **24.** digestive

Section 4.7

1. blood vessels; **2.** melanocytes; **3.** smooth; **4.** dermicidin; **5.** integumentary; **6.** epidermis; **7.** hypodermis; **8.** sebaceous; **9.** basal; **10.** squamous; **11.** papillae; **12.** keratin; **13.** root; **14.** shaft

Section 4.8

1. c; **2.** d; **3.** f; **4.** b; **5.** e; **6.** a; **7.** core temperature; **8.** skin, internal organs; **9.** hypothalamus; **10.** blood vessels, sweat glands; **11.** positive

Chapter Test

1. c; **2.** d; **3.** d; **4.** c; **5.** b; **6.** d; **7.** d; **8.** d; **9.** b; **10.** d; **11.** d; **12.** d; **13.** d; **14.** c; **15.** d; **16.** a; **17.** d; **18.** c; **19.** b; **20.** d; **21.** b; **22.** d; **23.** d; **24.** c; **25.** d

Chapter 5

Sections 5.1, 5.2, 5.3

1. c; **2.** f; **3.** i; **4.** m; **5.** h; **6.** j; **7.** a; **8.** k; **9.** e; **10.** l; **11.** b; **12.** g; **13.** d; **14.** support, movement, protection; **15.** formation of blood cells, mineral storage; **16.** collagen, elastin, water; **17.** fibrocartilage; **18.** hyaline; **19.** elastic; **20.** osteoblasts, osteoclasts; **21.** stress, osteoblasts; **22.** hematoma; **23.** callus; **24.** epiphysis; **25.** diaphysis; **26.** spongy bone; **27.** compact bone; **28.** yellow bone marrow; **29.** blood vessel; **30.** periosteum; **31.** central cavity; **32.** compact bone; **33.** spongy bone; **34.** osteon; **35.** blood vessels in central canal; **36.** hyaline; **37.** ossification; **38.** bone; **39.** chondroblasts; **40.** matrix; **41.** osteoblasts; **42.** osteoid; **43.** hydroxyapatite; **44.** osteocytes; **45.** lacunae; **46.** increases; **47.** one-third, two-thirds; **48.** stimulates; **49.** increase; **50.** increase; **51.** increases; **52.** increases

Section 5.4

1. carpals; **2.** metacarpals; **3.** phalanges; **4.** cranium; **5.** maxilla; **6.** mandible; **7.** clavicle; **8.** scapula; **9.** sternum; **10.** ribs; **11.** humerus; **12.** vertebrae; **13.** ulna; **14.** radius; **15.** coxal bone; **16.** sacrum; **17.** femur; **18.** patella; **19.** tibia; **20.** fibula; **21.** tarsals; **22.** metatarsals; **23.** phalanges; **24.** false; **25.** true; **26.** false; **27.** true; **28.** true

Section 5.5

1. a; **2.** b; **3.** a; **4.** c; **5.** a; **6.** a; **7.** b; **8.** a; **9.** a; **10.** c

Section 5.6

1. rheumatoid arthritis; **2.** sprain; **3.** bursitis; **4.** osteoarthritis

Chapter Test

1. c; **2.** b; **3.** d; **4.** c; **5.** a; **6.** a; **7.** d; **8.** c; **9.** a; **10.** b; **11.** c; **12.** c; **13.** c; **14.** e; **15.** c; **16.** b; **17.** a; **18.** b; **19.** b; **20.** c; **21.** d; **22.** d; **23.** c; **24.** c; **25.** a

Chapter 6

Sections 6.1, 6.2

1. d; **2.** f; **3.** j; **4.** h; **5.** a; **6.** i; **7.** l; **8.** k; **9.** e; **10.** b; **11.** g; **12.** c; **13.** produce, resist, generate; **14.** excitable; **15.** contract, relax; **16.** synergistic; **17.** antagonistic; **18.** insertion, origin; **19.** sarcomere; **20.** troponin-tropomyosin complex; **21.** bone; **22.** tendon; **23.** whole muscle; **24.** connective tissue sheath; **25.** fascicle; **26.** muscle fiber (cell); **27.** nuclei; **28.** myofibril; **29.** sarcomere; **30.** Z line; **31.** actin; **32.** myosin; **33.** acetylcholine; **34.** motor neuron; **35.** neurotransmitter; **36.** binding; **37.** electrical; **38.** T tubules; **39.** sarcoplasmic reticulum; **40.** troponin; **41.** troponin-tropomyosin complex; **42.** cross-bridges; **43.** bend; **44.** actin; **45.** sliding-filament; **46.** shortening; **47.** contraction; **48.** ATP; **49.** relaxation; **50.** sarcoplasmic reticulum; **51.** cross-bridges; **52.** lengthens; **53.** ATP; **54.** a; **55.** b; **56.** a; **57.** b; **58.** b

Section 6.3

1. c; **2.** d; **3.** e; **4.** a; **5.** f; **6.** b; **7.** b, c, e, h, i, k, n; **8.** a, d, f, g, j, l, m; **9.** it increases the amount of calcium present; **10.** stimulating a muscle so frequently that it has no chance to relax; **11.** strength and endurance; **12.** Strength training increases the strength of a particular muscle using resistance; this requires a quick burst of energy, so it targets fast-twitch fibers; **13.** Running is an aerobic activity requiring the contribution of slow-twitch fibers; slow-twitch fibers do not increase significantly in muscle mass; **14.** Muscle cells cannot increase in number; however, fast-twitch fibers can increase in muscle mass by increasing the number of myofibrils. Strength training, such as weight lifting, can increase muscle mass.

Sections 6.4, 6.5

1. b; **2.** a; **3.** b, c; **4.** a, b; **5.** c; **6.** b, c; **7.** a; **8.** a; **9.** a) inherited genetic disorder resulting in the absence of a membrane protein, b) loss of muscle fibers and muscle wasting; **10.** a) bacterial infection, b) tetanic contractions especially affecting the jaw and neck; **11.** a) ATP depletion, dehydration, ion imbalance, lactic acid accumulation, b) uncontrolled muscle contractions, pain; **12.** a) excessive muscle stretching, resulting in torn fibers, b) internal bleeding, swelling, pain; **13.** a) straining or tearing of the fascia around a muscle resulting in inflammation, b) pain

Chapter Test

1. a; **2.** b; **3.** c; **4.** d; **5.** c; **6.** b; **7.** a; **8.** e; **9.** a; **10.** a; **11.** b; **12.** a; **13.** b; **14.** d; **15.** b; **16.** b; **17.** a; **18.** b; **19.** b; **20.** c; **21.** c; **22.** b; **23.** d; **24.** a; **25.** c

Chapter 7

Section 7.1

1. o; **2.** e; **3.** i; **4.** k; **5.** f; **6.** r; **7.** t; **8.** a; **9.** p; **10.** b; **11.** g; **12.** s; **13.** c; **14.** d; **15.** m; **16.** h; **17.** j; **18.** l; **19.** q; **20.** n; **21.** red blood cells, white blood cells, platelets; **22.** O_2, CO_2; **23.** nucleus, organelles; **24.** iron atom; **25.** 120 days; **26.** liver, speen; **27.** defending; **28.** megakaryocytes; **29.** thrombopoietin; **30.** neutrophils; **31.** eosinophils; **32.** histamine, basophils; **33.** monocytes; **34.** lymphocytes; **35.** 9 days; **36.** liver, spleen; **37.** megakaryocyte, thrompopoietin; **38.** no; **39.** protein; **40.** high; **41.** lower than; **42.** increases; **43.** different sites; **44.** red blood cells; **45.** shorter; **46.** red blood cells, white blood cells; **47.** white blood cells; **48.** myeloblast; **49.** granular; **50.** first cells to combat infection, they engulf and destroy foreign cells; **51.** myeloblast; **52.** granular; **53.** defend the body against large parasites and release chemicals that affect allergic reactions; **54.** myeloblast; **55.** granular; **56.** secrete histamine; **57.** monoblast; **58.** agranular; **59.** differentiate into macrophages and stimulate lymphocytes; **60.** lymphoblast; **61.** agranular; **62.** critical to the body's immune system; **63.** kidney; **64.** erythropoietin; **65.** erythropoietin; **66.** red blood cells; **67.** bone marrow; **68.** red blood cells

Section 7.2

1. hemostasis; **2.** vascular spasm; **3.** contract; **4.** platelet plug; **5.** platelets; **6.** vessel; **7.** coagulation; **8.** clotting factors; **9.** prothrombin activator; **10.** prothrombin; **11.** thrombin; **12.** fibrinogen; **13.** fibrin; **14.** clot; **15.** hemophilia

Chapter 7, continued

Section 7.3

1. blood type; **2.** proteins; **3.** antigens; **4.** antibodies, lymphocytes; **5.** blood, lymph; **6.** antigen-antibody complex; **7.** antibodies; **8.** antibodies, agglutinate; **9.** organ damage, death; **10.** transfusion reaction; **11.** cross-matching; **12.** B, A; **13.** neither A nor B, A and B; **14.** type A+ also has Rh antigens; **15.** antibodies are produced only after exposure; **16.** a) Rh antibodies spread from mother to the fetus and attack fetal red blood cells, b) Mother: Rh−, Father: Rh+, c) no maternal antibodies are present yet, d) maternal injection of RhoGAM; **17.** samples of donor blood are mixed with recipient plasma and samples of recipient blood are mixed with donor plasma, absence of agglutination in both combinations

Section 7.4

1. d; **2.** a; **3.** e; **4.** f; **5.** b; **6.** g; **7.** c

Chapter Test

1. d; **2.** c; **3.** c; **4.** c; **5.** c; **6.** a; **7.** c; **8.** c; **9.** d; **10.** c; **11.** d; **12.** b; **13.** d; **14.** b; **15.** c; **16.** c; **17.** d; **18.** d; **19.** c; **20.** c; **21.** b; **22.** d; **23.** d; **24.** c; **25.** c

Chapter 8

Sections 8.1, 8.2

1. p; **2.** v; **3.** r; **4.** a; **5.** t; **6.** j; **7.** b; **8.** g; **9.** e; **10.** o; **11.** x; **12.** s; **13.** c; **14.** w; **15.** d; **16.** n; **17.** u; **18.** m; **19.** h; **20.** f; **21.** l; **22.** k; **23.** i; **24.** q; **25.** arteries, capillaries, veins; **26.** stretch; **27.** muscle; **28.** lumen; **29.** endothelium; **30.** smooth muscle; **31.** pores, slits; **32.** aneurysm; **33.** vasoconstriction; **34.** capillaries; **35.** single file; **36.** capillary beds; **37.** blood pressure; **38.** lymphatic capillaries; **39.** lymph; **40.** veins; **41.** varicose; **42.** pericardium; **43.** arrhythmia; **44.** atherosclerosis; **45.** blood transport, blood storage, regulation of blood flow into the capillaries; **46.** capillary walls consist of a single layer of endothelium comprised of epithelial cells; **47.** low pressure in the veins; **48.** skeletal muscle squeezes the veins, valves prevent blood from flowing backward, pressure associated with breathing pushes blood toward the heart; **49.** sinoatrial node, atrioventricular node, artrioventricular bundle, Purkinje fibers; **50.** atria; **51.** ventricles; **52.** systole; **53.** diastole; **54.** diastole; **55.** atria; **56.** ventricles; **57.** atrial systole; **58.** ventricular systole; **59.** AV; **60.** pulmonary trunk; **61.** aorta; **62.** systole; **63.** diastole; **64.** pulse; **65.** heart sounds; **66.** conduction; **67.** SA node; **68.** atrium; **69.** AV node; **70.** AV bundle; **71.** brain; **72.** a) right atrioventricular valve, b) right ventricle, c) pulmonary semilunar valve, d) pulmonary trunk, e) lungs, f) pulmonary veins, g) left atrium, h) left atrioventricular valve, i) left ventricle, j) aortic semilunar valve, k) aorta, l) arteries, m) arterioles, n) capillaries, o) venules, p) veins; **73.** a) aorta, b) pulmonary semilunar valve, c) right atrium, d) right AV valve, e) right ventricle, f) pulmonary trunk, g) left atrium, h) aortic semilunar valve, i) left AV valve, j) left ventricle, k) septum; **74.** c, d, h; **75.** a, f, g; **76.** b, e, i

Sections 8.3, 8.4, 8.5, 8.6

1. i; **2.** k; **3.** c; **4.** h; **5.** a; **6.** d; **7.** j; **8.** b; **9.** f; **10.** g; **11.** e; **12.** sphygmomanometer; **13.** systolic; **14.** isolated systolic hypertension; **15.** hypotension; **16.** heart rate, stroke volume; **17.** sympathetic, medulla oblongata; **18.** one; **19.** angiogram; **20.** CABG; **21.** a) systolic pressure,

b) ventricular contraction; **22.** a) diastolic pressure, b) ventricular relaxation; **23.** greater distance from the heart causes blood pressure to dissipate; **24.** blood pressure is steady in the capillaries and pulsatile in the arteries; **25.** blood moves with the presure gradient; **26.** arterial; **27.** aorta; **28.** carotid arteries; **29.** brain; **30.** heart rate; **31.** increase; **32.** negative feedback; **33.** medulla oblongata; **34.** sympathetic; **35.** parasympathetic; **36.** diameter; **37.** epinephrine; **38.** adrenal gland; **39.** oxygen; **40.** carbon dioxide; **41.** dilate

Chapter Test

1. increases, increasing; **2.** thinner; **3.** larger; **4.** less than; **5.** less than; **6.** increases; **7.** deoxygenated; **8.** d; **9.** c; **10.** a; **11.** b; **12.** c; **13.** c; **14.** c; **15.** b; **16.** a; **17.** a; **18.** a; **19.** b; **20.** c; **21.** d; **22.** a; **23.** d; **24.** c; **25.** c

Chapter 9

Sections 9.1, 9.2, 9.3, 9.4, 9.5

1. g; **2.** r; **3.** h; **4.** q; **5.** l; **6.** p; **7.** a; **8.** m; **9.** o; **10.** e; **11.** i; **12.** k; **13.** d; **14.** u; **15.** f; **16.** t; **17.** n; **18.** b; **19.** j; **20.** s; **21.** c; **22.** capillaries; **23.** cardiovascular; **24.** cells; **25.** fluid; **26.** vessels; **27.** nodes; **28.** neck; **29.** cardiovascular; **30.** macrophages; **31.** lymphocytes; **32.** lymph; **33.** lymphatic; **34.** thoracic; **35.** subclavian; **36.** a) thymus gland, b) aids maturation of T cells; **37.** a) lymph node, b) filters and cleanses lymph; **38.** a) lymph vessels, b) transport lymph; **39.** a) adenoids, b) protect the throat; **40.** a) tonsils, b) protect the throat; **41.** a) red pulp, b) contains macrophages that break down microorganisms, old red blood cells, and platelets; **42.** a) white pulp, b) contains lymphocytes that fight infection; **43.** a) spleen, b) removes damaged red blood cells and helps to fight infections

Sections 9.6, 9.7, 9.8, 9.9

1. (across) cytotoxic; **1.** (down) cytokines **2.** presenting; **3.** rejection; **4.** thymus; **5.** primary; **6.** bacteria; **7.** plasma; **8.** antibody; **9.** monoclonal; **10.** vaccine; **11.** (across) major **11.** (down) memory; **12.** variable; **13.** perforin; **14.** helper; **15.** f; **16.** c; **17.** g; **18.** i; **19.** a; **20.** e; **21.** h; **22.** d; **23.** b; **24.** j; **25.** k; **26.** IgA; **27.** 15%; **28.** IgE; **29.** 0.1%; **30.** IgG; **31.** 75%; **32.** IgD; **33.** less than 1%; **34.** IgM; **35.** 5–10%

Sections 9.10, 9.11

1. allergy; **2.** systemic; **3.** autoimmune; **4.** II; **5.** g; **6.** b; **7.** d; **8.** a; **9.** e; **10.** c; **11.** f

Chapter Test

1. d; **2.** b; **3.** a; **4.** b; **5.** d; **6.** c; **7.** b; **8.** a; **9.** c; **10.** a; **11.** c; **12.** a; **13.** b; **14.** c; **15.** b; **16.** a; **17.** b; **18.** b; **19.** b; **20.** b; **21.** a; **22.** b; **23.** a; **24.** c; **25.** c

Chapter 10

Sections 10.1, 10.2

1. f; **2.** h; **3.** k; **4.** n; **5.** l; **6.** o; **7.** b; **8.** j; **9.** m; **10.** g; **11.** a; **12.** d; **13.** e; **14.** i; **15.** c; **16.** respiratory; **17.** internal; **18.** cellular; **19.** sound; **20.** nose, pharynx; **21.** larynx, trachea, bronchi, lungs; **22.** upper; **23.** diffusion; **24.** pharynx; **25.** epiglottis, larynx; **26.** trachea, connective tissue, muscle; **27.** bronchi; **28.** bronchioles; **29.** fluid, pleural cavity; **30.** three, two; **31.** one, squamous epithelial; **32.** surfactant; **33.** alveoli, pulmonary capillaries; **34.** sinus; **35.** nasal cavity; **36.** external nose; **37.** nostril; **38.** tongue; **39.** larynx; **40.** esophagus; **41.** trachea; **42.** sinus; **43.** opening of eustachian tube; **44.** pharynx; **45.** glottis; **46.** epiglottis

Section 10.3

1. collide; **2.** decreases; **3.** high, low; **4.** pressure gradient; **5.** intercostals, diaphragm; **6.** intercostals lift the ribs upward, diaphragm flattens; **7.** increases; **8.** decreases; **9.** air pressure outside the body is greater than air pressure in the lungs; **10.** decreases; **11.** increases; **12.** air pressure in the lungs is greater than air pressure outside the body; **13.** tidal volume; **14.** vital capacity; **15.** inspiratory reserve volume; **16.** expiratory reserve volume; **17.** residual volume

Section 10.4

1. a) 104, b) 100, c) into the pulmonary capillaries, d) 40, e) 46, f) into the alveoli, g) external; **2.** a) 100, b) <40, c) into the cell, d) 46, e) >46, f) into the capillary, g) internal; **3.** a) hemoglobin, oxygen, oxyhemoglobin, b) reversible, c) neutral pH, cool temperature, high PO_2 in the lungs; **4.** a) hemoglobin, carbon dioxide, carbaminohemoglobin, b) about 20%; **5.** a) bicarbonate, b) about 70%

Sections 10.5, 10.6

1. g; **2.** e; **3.** i; **4.** b; **5.** d; **6.** k; **7.** m; **8.** a; **9.** j; **10.** f; **11.** c; **12.** h; **13.** l

Chapter Test

1. c; **2.** c; **3.** b; **4.** b; **5.** d; **6.** d; **7.** a; **8.** a; **9.** c; **10.** c; **11.** a; **12.** a; **13.** c; **14.** b; **15.** b; **16.** d; **17.** a; **18.** a; **19.** d; **20.** d; **21.** a; **22.** c; **23.** c; **24.** a; **25.** c

Chapter 11

Sections 11.1, 11.2, 11.3, 11.4

1. d; **2.** g; **3.** n; **4.** j; **5.** b; **6.** q; **7.** a; **8.** m; **9.** k; **10.** o; **11.** h; **12.** f; **13.** p; **14.** c; **15.** i; **16.** l; **17.** e; **18.** energy, transmission, axons; **19.** oligodendrocytes; **20.** sodium-potassium; **21.** −70 mV, negative; **22.** voltage-; **23.** three, two; **24.** all-or-none; **25.** excitatory; **26.** presynaptic membrane; **27.** postsynaptic membrane; **28.** inhibitory; **29.** convergence; **30.** b, m; **31.** c, o; **32.** a, j; **33.** g, i; **34.** e, k; **35.** d, l; **36.** f, p; **37.** h, n; **38.** given as an example; **39.** a; **40.** hyperpolarizing graded potential; **41.** c; **42.** repolarization phase of an action potential; **43.** g; **44.** action potential; **45.** f; **46.** threshold; **47.** h; **48.** depolarizing graded potential; **49.** b; **50.** depolarization phase of an action potential; **51.** e; **52.** summation; **53.** d; **54.** are not; **55.** into; **56.** is; **57.** increases; **58.** open, into; **59.** short-lived; **60.** do not, one; **61.** greater

Sections 11.5, 11.6

1. (across) postganglionic; **1.** (down) parasympathetic; **2.** preganglionic; **3.** sympathetic; **4.** autonomic; **5.** two; **6.** somatic; **7.** meninges; **8.** white; **9.** ventral; **10.** ventricle; **11.** ganglia; **12.** nerves; **13.** cranial; **14.** dorsal; **15.** gray; **16.** reflex; **17.** spinal

Sections 11.7, 11.8, 11.9, 11.10

1. f; **2.** d; **3.** b; **4.** g; **5.** i; **6.** a; **7.** c; **8.** h; **9.** e; **10.** c, l; **11.** b, k; **12.** f, j; **13.** d, h; **14.** a, i; **15.** e, g

Sections 11.11, 11.12

1. a drug that affects higher brain function; **2.** they cross the blood-brain barrier; **3.** they influence the concentration or mode of action of brain neurotransmitters; **4.** the liver detoxifies drugs; rising levels of liver enzymes necessitate higher dosages to achieve the same effect; **5.** the need to continue obtaining and using a substance despite one's better judgment and good intentions; **6.** concussion; **7.** spinal cord injury; **8.** encephalitis; **9.** meningitis; **10.** rabies; **11.** epilepsy; **12.** Parkinson's disease; **13.** Alzheimer's disease; **14.** brain tumor

Chapter Test

1. b; **2.** a; **3.** c; **4.** c; **5.** d; **6.** c; **7.** b; **8.** a; **9.** c; **10.** d; **11.** a; **12.** d; **13.** a; **14.** d; **15.** b; **16.** b; **17.** d; **18.** b; **19.** c; **20.** a; **21.** d; **22.** a; **23.** c; **24.** b; **25.** c

Chapter 12

Sections 12.1, 12.2, 12.3

1. e; **2.** h; **3.** j; **4.** k; **5.** a; **6.** m; **7.** o; **8.** l; **9.** b; **10.** i; **11.** g; **12.** n; **13.** d; **14.** f; **15.** c; **16.** not conscious; **17.** quickly; **18.** most; **19.** fast pain; **20.** action potential; **21.** sweet, sour, salty, bitter; **22.** tastants, odorants; **23.** olfactory hairs

Sections 12.4, 12.5

1. outer ear; **2.** middle ear; **3.** inner ear; **4.** pinna; **5.** malleus; **6.** incus; **7.** auditory canal; **8.** tympanic membrane; **9.** semicircular canals; **10.** vestibular nerve; **11.** cochlear nerve; **12.** vestibule; **13.** oval window; **14.** stapes; **15.** cochlea; **16.** round window; **17.** eustachian tube; **18.** waves; **19.** amplitude; **20.** frequency; **21.** pinna; **22.** eardrum; **23.** malleus; **24.** incus; **25.** stapes; **26.** oval window; **27.** vestibular; **28.** cochlea; **29.** basilar; **30.** tectorial; **31.** neurotransmitter; **32.** auditory; **33.** base; **34.** tip; **35.** round window; **36.** c; **37.** f; **38.** e; **39.** b; **40.** g; **41.** a; **42.** d

Sections 12.6, 12.7

1. Canal of Schlemm; **2.** iris; **3.** lens; **4.** pupil; **5.** cornea; **6.** aqueous humor; **7.** ciliary muscle; **8.** sclera; **9.** choroid; **10.** retina; **11.** fovea; **12.** optic disk; **13.** optic nerve; **14.** vitreous humor; **15.** covers and protects eyeball; **16.** focuses light; **17.** nourishes and cushions cornea and lens; **18.** adjusts amount of incoming light; **19.** regulates focus; **20.** adjusts curvature of the lens; **21.** transmits light to retina; **22.** absorbs light and converts it into impulses; **23.** rods, cones; **24.** photopigment; **25.** rhodopsin; **26.** cones; **27.** ganglion cells; **28.** retinal detachment; **29.** glaucoma; **30.** cataracts; **31.** otitis media; **32.** Ménière's syndrome; **33.** myopia

Chapter Test

1. c; **2.** a; **3.** b; **4.** d; **5.** a; **6.** a; **7.** a; **8.** c; **9.** b; **10.** b; **11.** d; **12.** c; **13.** d; **14.** c; **15.** a; **16.** a; **17.** d; **18.** c; **19.** d; **20.** a; **21.** b; **22.** a; **23.** c; **24.** c; **25.** b

Chapter 13

Sections 13.1, 13.2, 13.3

1. d; **2.** g; **3.** f; **4.** j; **5.** h; **6.** i; **7.** a; **8.** c; **9.** e; **10.** b; **11.** bloodstream, cells; **12.** 50; **13.** receptors; **14.** slower; **15.** endocrine, nervous; **16.** steroid; **17.** nonsteroid; **18.** target cells; **19.** hypothalamus, posterior pituitary; **20.** hormones, hypothalamus; **21.** neural, endocrine; **22.** ADH; **23.** acromegaly; **24.** hypothalamus—releasing and inhibiting hormones; **25.** pituitary gland—see #38 and #39; **26.** stomach—gastrin; **27.** pancreas—glucagons, insulin, somatostatin; **28.** kidneys—rennin, erythropoietin; **29.** ovaries—estrogen, progesterone; **30.** pineal gland—melatonin; **31.** thyroid gland—thyroxin, calcitonin; **32.** parathyroid gland—parathyroid hormone; **33.** thymus gland—thymosin, thymopoietin; **34.** heart—atrial natriuretic hormone; **35.** adrenal glands—cortex: cortisol, aldosterone—medulla: epinephrine, norepinephrine; **36.** intestines—secretin, cholecystokinin; **37.** testes—testosterone; **38.** anterior lobe of the pituitary—ACTH, TSH, FSH, LH, prolactin, growth hormone; **39.** posterior lobe of the pituitary—ADH, oxytocin; **40.** cholesterol; **41.** insoluble; **42.** not enter; **43.** intracellular; **44.** membrane ion channel; **45.** slower

Sections 13.4, 13.5, 13.6, 13.7, 13.8

1. f; **2.** h; **3.** m; **4.** j; **5.** l; **6.** a; **7.** n; **8.** d; **9.** p; **10.** g; **11.** o; **12.** c; **13.** q; **14.** e; **15.** i; **16.** b; **17.** k; **18.** parathyroid; **19.** aldosterone; **20.** calcitonin; **21.** glucagons; **22.** insulin; **23.** epinephrine; **24.** cortisol; **25.** thyroxin; **26.** atrial; **27.** TSH

Sections 13.9, 13.10

1. histamine; **2.** prostaglandins; **3.** nitric oxide; **4.** growth factors

Chapter Test

1. b; **2.** a; **3.** b; **4.** b; **5.** c; **6.** b; **7.** b; **8.** b; **9.** a; **10.** d; **11.** d; **12.** a; **13.** b; **14.** d; **15.** c; **16.** b; **17.** a; **18.** b; **19.** b; **20.** b; **21.** c; **22.** b; **23.** d; **24.** c; **25.** c

Chapter 14

Sections 14.1, 14.2, 14.3, 14.4, 14.5

1. e; **2.** j; **3.** q; **4.** h; **5.** o; **6.** r; **7.** a; **8.** m; **9.** n; **10.** p; **11.** b; **12.** g; **13.** i; **14.** d; **15.** k; **16.** c; **17.** l; **18.** f; **19.** g, m; **20.** d, q; **21.** l, t; **22.** b, v; **23.** c, n; **24.** e, o; **25.** h, w; **26.** a, p; **27.** i, u; **28.** j, r; **29.** f, x; **30.** k, s; **31.** lumen; **32.** smooth, circular, lengthwise; **33.** mouth; **34.** incisor; **35.** crown; **36.** amylase; **37.** involuntary; **38.** pyloric sphincter; **39.** gastric juice; **40.** peristalsis; **41.** duodenum; **42.** jejunum, ileum; **43.** microvilli; **44.** longitudinal layer of smooth muscle; **45.** circular layer of smooth muscle; **46.** lumen; **47.** mucosa; **48.** submucosa; **49.** muscularis; **50.** serosa; **51.** muscularis; **52.** submucosa; **53.** serosa; **54.** muscularis

Sections 14.6, 14.7, 14.8

1. hepatic portal system; **2.** liver; **3.** pancreas; **4.** colon; **5.** cecum; **6.** sigmoid colon; **7.** liver; **8.** bile; **9.** bile; **10.** bacteria; **11.** bile salts; **12.** fat globules; **13.** emulsification droplets; **14.** free fatty acids and monoglycerides; **15.** micelles; **16.** diffusion; **17.** amino acids; **18.** chylomicron; **19.** exocytosis; **20.** lacteal

Section 14.9

1. peristalsis, gastric juice; **2.** gastrin, gastric juice; **3.** secretin, bicarbonate; **4.** cholecystokinin (CCK), digestive; **5.** bile; **6.** inhibit, chyme; **7.** motility

Sections 14.10, 14.11, 14.12

1. h; **2.** p; **3.** l; **4.** g; **5.** q; **6.** u; **7.** a; **8.** t; **9.** e; **10.** o; **11.** n; **12.** d; **13.** r; **14.** b; **15.** j; **16.** s; **17.** f; **18.** m; **19.** c; **20.** k; **21.** i

Chapter Test

1. c; **2.** b; **3.** c; **4.** d; **5.** a; **6.** d; **7.** b; **8.** b; **9.** a; **10.** c; **11.** a; **12.** d; **13.** d; **14.** d; **15.** c; **16.** d; **17.** a; **18.** c; **19.** c; **20.** c; **21.** c; **22.** d; **23.** c; **24.** d; **25.** a

Chapter 15

Sections 15.1, 15.2, 15.3

1. g; **2.** p; **3.** m; **4.** k; **5.** b; **6.** r; **7.** n; **8.** q; **9.** a; **10.** l; **11.** o; **12.** e; **13.** h; **14.** f; **15.** j; **16.** c; **17.** i; **18.** d; **19.** kidney; **20.** major blood vessels; **21.** ureter; **22.** bladder; **23.** urethra; **24.** renal blood vessels; **25.** nephron; **26.** renal pelvis; **27.** cortex; **28.** medulla; **29.** ureter; **30.** efferent arterioles; **31.** glomerulus; **32.** peritubular capillaries; **33.** artery; **34.** vein; **35.** descending limb; **36.** ascending limb; **37.** loop of Henle; **38.** collecting duct; **39.** distal tubule; **40.** afferent arteriole; **41.** glomerular capsule; **42.** proximal tubule

Sections 15.4, 15.5, 15.6

1. glomerular filtration, tubular reabsorption, tubular secretion; **2.** glomerular capillaries, glomerular capsule; **3.** solutes, proteins, cells; **4.** blood pressure; **5.** tubules, peritubular capillaries; **6.** proximal; **7.** glucose, bicarbonate, water, sodium; **8.** 10%, regulated; **9.** active transport; **10.** capillaries, tubule, excreted; **11.** H^+, NH_4^+; **12.** K^+; **13.** less; **14.** concentration, renal medulla; **15.** urea; **16.** reabsorb; **17.** ADH; **18.** increases, collecting duct; **19.** water, concentrated; **20.** micturition reflex; **21.** smooth, skeletal; **22.** ascending limb of the loop of Henle; **23.** proximal tubule; **24.** upper portion of ascending limb; **25.** ascending limb; **26.** lower portion of ascending limb; **27.** collecting duct; **28.** loop of Henle; **29.** vasa recta

Sections 15.7, 15.8

1. a) renin, b) angiotensinogen, c) angiotensin I, d) vasoconstriction, e) blood pressure, f) aldosterone, g) salt, h) ADH, i) water; **2.** increased concentration of blood solutes; **3.** increased reabsorption of Na^+; **4.** decreased blood volume; **5.** afferent arteriole cells of the juxtaglomerular apparatus; **6.** heart; **7.** inhibits sodium reabsorption in the distal tubule and collecting duct; **8.** renin and aldosterone; **9.** H^+; **10.** ammonia; **11.** bicarbonate; **12.** H^+; **13.** excreted; **14.** erythropoietin; **15.** decrease; **16.** increase; **17.** decreases; **18.** Vitamin D; **19.** calcium; **20.** phosphorus; **21.** UV rays; **22.** liver; **23.** kidneys; **24.** kidney stones; **25.** acute renal failure; **26.** chronic renal failure; **27.** hemodialysis; **28.** continuous ambulatory peritoneal dialysis (CAPD)

Chapter Test

1. b; **2.** a; **3.** d; **4.** a; **5.** c; **6.** c; **7.** d; **8.** d; **9.** b; **10.** a; **11.** c; **12.** c; **13.** c; **14.** b; **15.** b; **16.** b; **17.** a; **18.** c; **19.** a; **20.** b; **21.** a; **22.** b; **23.** d; **24.** b; **25.** a

Chapter 16

Section 16.1

1. f; 2. h; 3. n; 4. q; 5. l; 6. k; 7. p; 8. b; 9. c; 10. a; 11. o; 12. g; 13. i; 14. e; 15. j; 16. m; 17. d; 18. fructose, prostaglandins; 19. spermatogonia; 20. 46, 23; 21. inhibin; 22. mitosis; 23. spermatids; 24. midpiece; 25. negative, pituitary gland; 26. GnRH; 27. LH; 28. vas deferens; 29. epididymis; 30. testis; 31. penis; 32. seminal vesicle; 33. prostate gland; 34. bulbourethral gland; 35. urethra; 36. seminiferous tubules

Section 16.2

1. e; 2. j; 3. g; 4. l; 5. i; 6. k; 7. m; 8. d; 9. a; 10. h; 11. n; 12. c; 13. f; 14. b; 15. oviduct; 16. endometrium; 17. areola; 18. mammary; 19. prolactin, oxytocin; 20. oviduct; 21. fimbriae; 22. ovary; 23. uterus; 24. bladder; 25. urethra; 26. clitoris; 27. labium minora; 28. labium majora; 29. cervix; 30. vagina

Sections 16.3, 16.4, 16.5

1. ovarian; 2. primary; 3. Meiosis I; 4. puberty; 5. Meiosis II; 6. granulose; 7. follicle; 8. FSH; 9. anterior; 10. pituitary; 11. GnRH; 12. hypothalamus; 13. ovulation; 14. glycoprotein; 15. zona; 16. pellucida; 17. antrum; 18. progesterone; 19. Meiosis I; 20. secondary; 21. Graafian; 22. LH; 23. ovulation; 24. corpus; 25. luteum; 26. endometrium; 27. decrease; 28. ovarian; 29. days 1–5; 30. estrogen and progesterone levels decline; 31. days 6–14; 32. rising levels of estrogen and progesterone; 33. ovulation; 34. secretory phase; 35. progesterone; 36. implantation of a fertilized egg; 37. degeneration of the corpus luteum; 38. before; 39. rising levels of estrogen in the first half of the proliferative phase trigger a surge in LH; 40. high levels of progesterone and estrogen, secreted by the corpus luteum, feed back to the hypothalamus and inhibit FSH secretion; 41. vasectomy; 42. Depo-Provera; 43. spermicide; 44. intrauterine device (IUD); 45. rhythm method; 46. Lunelle; 47. Ortho Evra; 48. Mifeprex

Section 16.6, 16.7

1. k; 2. e; 3. f; 4. j; 5. h; 6. i; 7. a; 8. d; 9. g; 10. b; 11. c

Chapter Test

1. d; 2. d; 3. a; 4. c; 5. a; 6. c; 7. b; 8. a; 9. b; 10. a; 11. c; 12. c; 13. d; 14. d; 15. c; 16. c; 17. b; 18. a; 19. a; 20. c; 21. c; 22. c; 23. d; 24. d; 25. c

Chapter 17

Sections 17.1, 17.2

1. e; 2. q; 3. m 4. p; 5. n; 6. s; 7. a; 8. r; 9. g; 10. b; 11. o; 12. i; 13. k; 14. h; 15. j; 16. c; 17. d; 18. l; 19. f; 20. cytokinesis; 21. arrested; 22. G2; 23. chromatin; 24. histones; 25. DNA polymerase; 26. repair; 27. bubbles; 28. uracil; 29. promotor; 30. TAGCCT; 31. Ile; 32. c, e, h, j; 33. a, f, i, l; 34. b, d, g, k; 35. TGATGGGTACGGTC; 36. UGAUGGGUACGGUC; 37. met-gly-thr-val; 38. c; 39. a; 40. l; 41. h; 42. f; 43. k; 44. i; 45. d; 46. e; 47. b; 48. g; 49. j; 50. o; 51. m; 52. n

Sections 17.3, 17.4, 17.5

1. e; **2.** i; **3.** m; **4.** k; **5.** o; **6.** a; **7.** n; **8.** c; **9.** l; **10.** f; **11.** h; **12.** g; **13.** d; **14.** j; **15.** b; **16.** Meiosis I; **17.** Meiosis I; **18.** Mitosis; **19.** Meiosis II; **20.** Mitosis; **21.** Meiosis I; **22.** Meiosis II; **23.** d; **24.** b; **25.** f; **26.** g; **27.** e; **28.** c; **29.** m; **30.** l; **31.** a; **32.** k; **33.** i; **34.** h; **35.** j; **36.** 46; **37.** a prophase cell has twice the amount of DNA because the chromosomes are replicated; each chromosome consists of two pieces of DNA; **38.** yes; **39.** chromatin; **40.** false; **41.** true; **42.** true; **43.** false; **44.** true; **45.** true; **46.** true; **47.** true

Chapter Test

1. b; **2.** c; **3.** a; **4.** c; **5.** d; **6.** c; **7.** b; **8.** c; **9.** c; **10.** b; **11.** c; **12.** a; **13.** b; **14.** c; **15.** b; **16.** b; **17.** d; **18.** c; **19.** c; **20.** d; **21.** a; **22.** d; **23.** b; **24.** d; **25.** d

Chapter 18

Sections 18.1, 18.2, 18.3

1. e; **2.** g; **3.** i; **4.** j; **5.** l; **6.** a; **7.** m; **8.** k; **9.** c; **10.** h; **11.** d; **12.** b; **13.** f; **14.** dysplasia; **15.** four; **16.** growth; **17.** proteins; **18.** multiple; **19.** inhibit; **20.** hepatitis B; **21.** initiators; **22.** promoters; **23.** 30; **24.** 30; **25.** free radicals; **26.** A, C, E; **27.** immune; **28.** e; **29.** c; **30.** a, h; **31.** f, g; **32.** b, i; **33.** d

Sections 18.4, 18.5

1. PET scan; **2.** genetic testing; **3.** enzyme testing; **4.** chemotherapy; **5.** photodynamic therapy; **6.** gene therapy; **7.** "starving" cancer; **8.** cancer cells divide rapidly, while most normal body cells do not undergo continuous cell division; **9.** the cancer antigen is used to produce antibodies that will target the cancer cell for destruction by the immune system

Chapter Test

1. c; **2.** c; **3.** b; **4.** d; **5.** b; **6.** c; **7.** b; **8.** b; **9.** c; **10.** c; **11.** c; **12.** c; **13.** b; **14.** d; **15.** c; **16.** c; **17.** c; **18.** c; **19.** c; **20.** c; **21.** d; **22.** b; **23.** a; **24.** a; **25.** d

Chapter 19

Sections 19.1, 19.2

1. e; **2.** l; **3.** n; **4.** k; **5.** p; **6.** m; **7.** q; **8.** a; **9.** o; **10.** f; **11.** h; **12.** d; **13.** g; **14.** b; **15.** i; **16.** r; **17.** j; **18.** c; **19.** genome; **20.** Gregor Mendel; **21.** independent assortment; **22.** genotype, environment; **23.** linkage group; **24.** genes; **25.** Aa; **26.** Aa; **27.** AA; **28.** Aa; **29.** Aa; **30.** aa; **31.** a) Tt, b) tt, c) TT, d)

	T	T
t	Tt	Tt
t	Tt	Tt

e) 50%;

32. a) incomplete dominance, b)

	W	W
w	Ww	Ww
w	Ww	Ww

c) 100%; **33.** a) OO, b) Type AB, c)

	A	B
O	AO	BO
O	AO	BO

Chapter 19, continued

Sections 19.3, 19.4, 19.5, 19.6

1. karyotype; 2. autosomes, sex; 3. Y; 4. sex-linked; 5. sex-influenced; 6. deletion; 7. translocation; 8. dominant; 9. males inherit only one X chromosome; they cannot be protected by a normal allele on a second X chromosome; 10. men receive a Y chromosome from their father; their mother is the only parent that donates an X chromosome; 11. the father has only one X chromosome—the one with the trait—and he will pass it to all of his daughters;
12. a) X^cX^c, b) X^CX^c, c) X^cY, d)

	X^C	Y
X^c	X^CX^c	X^cY
X^c	X^CX^c	X^cY

e) 100%, f) 0%;

13. proteins play many roles in body chemistry that could influence behavior; 14. Turner syndrome; 15. PKU; 16. cri-du-chat syndrome; 17. Klinefelter syndrome; 18. Down syndrome

Chapter Test

1. d; 2. b; 3. a; 4. c; 5. c; 6. b; 7. c; 8. b; 9. d; 10. c; 11. c; 12. c; 13. a; 14. c; 15. d; 16. b; 17. d; 18. c; 19. b; 20. a; 21. d; 22. b; 23. d; 24. b; 25. a

---

Chapter 20

Sections 20.1, 20.2

1. f; 2. i; 3. j; 4. c; 5. h; 6. e; 7. b; 8. a; 9. g; 10. d; 11. base; 12. test tube; 13. primers; 14. strands; 15. synthesis; 16. labeled; 17. fluorescent; 18. synthesis; 19. polymerase; 20. lengths; 21. labeled; 22. gel electrophoresis; 23. size; 24. smaller; 25. larger; 26. fluoroscope; 27. complementary; 28. original; 29. base, restriction; 30. bacteria; 31. reproduction; 32. DNA; 33. regulatory; 34. heating; 35. two; 36. proteins; 37. unwind; 38. d; 39. i; 40. f; 41. g; 42. c; 43. a; 44. e; 45. b; 46. h

Sections 20.3, 20.4

1. retroviruses; 2. hepatitis B; 3. fertilized; 4. engineered; 5. failures; 6. ten; 7. bacteria; 8. transgenic; 9. vector; 10. immuno; 11. interleukins; 12. Genome; 13. liposomes; 14. difficult; 15. therapy; 16. proteins; 17. nasal; 18. pharming; 19. vaccines

Chapter Test

1. d; 2. c; 3. d; 4. d; 5. b; 6. a; 7. b; 8. a; 9. c; 10. b; 11. b; 12. b; 13. d; 14. a; 15. d; 16. d; 17. d; 18. d; 19. a; 20. d; 21. d; 22. c; 23. b; 24. d; 25. b

---

Chapter 21

Sections 21.1, 21.2

1. acrosome; 2. head; 3. midpiece; 4. tail; 5. corona radiata; 6. cytoplasm; 7. nucleus; 8. plasma membrane; 9. first polar body; 10. zona pellucida; 11. acrosome; 12. upper third of the oviduct; 13. vagina, cervix, uterus, oviduct; 14. 24 hours; 15. zona pellucida and corona radiata; 16. entry of the sperm's nucleus into the egg; 17. sperm entry triggers the release of enzymes that cause changes in the zona pellucida, making it impenetrable to additional sperm; 18. entry of the sperm; 19. zygote; 20. when the nuclei of the sperm and mature ovum unite;

21. follicle cells and polar bodies; **22.** e; **23.** d; **24.** h; **25.** g; **26.** a; **27.** c; **28.** f; **29.** b

Sections 21.3, 21.4

1. j; **2.** d; **3.** h; **4.** l; **5.** g; **6.** o; **7.** b; **8.** a; **9.** f; **10.** c; **11.** i; **12.** m; **13.** k; **14.** n; **15.** q; **16.** u; **17.** t; **18.** p; **19.** e; **20.** r; **21.** s; **22.** 2 weeks; **23.** morula; **24.** morula; **25.** embryonic disc; **26.** blastocyst; **27.** implantation; **28.** trophoblast cells; **29.** the appearance of an amniotic cavity, and of ectoderm and endoderm in the embryonic disc; **30.** week 3 through week 8; **31.** the appearance of mesoderm; **32.** mesoderm; **33.** endoderm; **34.** ectoderm; **35.** placenta; **36.** yolk sac; **37.** amniotic cavity; **38.** umbilical cord; **39.** uterus; **40.** amnion; **41.** embryonic; **42.** amnion; **43.** allantois; **44.** yolk sac; **45.** chorion; **46.** amnion; **47.** umbilical cord; **48.** digestive; **49.** gonads; **50.** trophoblast; **51.** placenta; **52.** chorion; **53.** endometrium; **54.** umbilical cord; **55.** endometrium; **56.** villi

Sections 21.5, 21.6, 21.7, 21.8, 21.9

1–5: See textbook; **6.** pituitary gland; **7.** positive; **8.** dilation, expulsion, afterbirth; **9.** episiotomy; **10.** ductus venosus; **11.** foramen ovale; **12.** connective; **13.** neonatal; **14.** childhood; **15.** GnRH; **16.** childhood; **17.** telomere; **18.** LH, FSH; **19.** a) less elastic, b) stiffen, c) decreases, d) decreases, e) stiffens, f) decreases, g) decreases, h) enlarges; **20.** irreversible

Chapter Test

1. c; **2.** d; **3.** c; **4.** a; **5.** b; **6.** c; **7.** a; **8.** a; **9.** b; **10.** a; **11.** b; **12.** c; **13.** a; **14.** a; **15.** c; **16.** a; **17.** b; **18.** d; **19.** a; **20.** b; **21.** c; **22.** b; **23.** b; **24.** a.; **25.** c

Chapter 22

Sections 22.1, 22.2

1. g; **2.** n; **3.** j; **4.** k; **5.** m; **6.** a; **7.** o; **8.** l; **9.** d; **10.** e; **11.** c; **12.** h; **13.** f; **14.** b; **15.** i; **16.** descent with modification; **17.** sediment, volcanic ash; **18.** stratification; **19.** lowest; **20.** potassium; **21.** pharyngeal pouches; **22.** biochemical molecules; **23.** diverge; **24.** biogeography; **25.** Pangaea; **26.** mass extinction; **27.** bottleneck effect; **28.** a) homologous, b) descent from a common ancestor, c) divergence allowed each species to evolve along a different path; **29.** a) analogous, b) no, c) they evolved to carry out a similar function

Sections 22.3, 22.4, 22.5

1. b; **2.** i; **3.** f; **4.** k; **5.** h; **6.** a; **7.** d; **8.** j; **9.** c; **10.** e; **11.** g

Section 22.6

1. Animalia; **2.** eukaryotic, multicellular, complex anatomy; **3.** Chordata; **4.** nerve cord, backbone; **5.** Mammalia; **6.** mammary glands; **7.** Primata; **8.** five digits, flat fingernails; **9.** Hominidae **10.** upright posture, enlarged brain; **11.** Homo; **12.** human form; **13.** Sapiens; **14.** complex social structure, complex spoken language; **15.** g; **16.** a; **17.** h; **18.** i; **19.** b; **20.** e; **21.** f; **22.** d; **23.** c; **24.** 4–5 million years ago; **25.** *Australopithecus anamensis;* **26.** upright posture; **27.** 100,000 to 140,000 years ago; **28.** *Homo habilis;* **29.** *Homo ergaster;* **30.** *Homo heidelbergensis;* **31.** 35,000 years; **32.** increased brain size, development of spoken language, development of a human physical structure

Chapter Test

1. a; **2.** c; **3.** b; **4.** c; **5.** b; **6.** b; **7.** a; **8.** a; **9.** c; **10.** c; **11.** c; **12.** d; **13.** a; **14.** b; **15.** b; **16.** c; **17.** b; **18.** b; **19.** b; **20.** b; **21.** b; **22.** d; **23.** c; **24.** d; **25.** c

Chapter 23

Sections 23.1, 23.2, 23.3

1. h; 2. k; 3. f; 4. i; 5. c; 6. l; 7. j; 8. a; 9. m; 10. n; 11. g; 12. e; 13. b; 14. d; 15. habitat; 16. competition for resources, intolerable conditions, physical obstacles; 17. biotic potential, carrying capacity; 18. environmental resistance; 19. increase; 20. competitive exclusion; 21. climax communities, complexity; 22. chemical elements, energy; 23. under ideal conditions the population will double repeatedly; 24. the effect of environmental resistance; 25. a change in environmental conditions; 26. the number of offspring produced by each reproducing member, the length of time required for offspring to reach sexual maturity, the ratio of males to females, the number of reproductive members of the population

Sections 23.4, 23.5

1. l; 2. e; 3. g; 4. m; 5. j; 6. h; 7. f; 8. a; 9. i; 10. d; 11. c; 12. k; 13. b; 14. c; 15. a; 16. d; 17. b; 18. f, j, k, l; 19. f, j, k; 20. g, k; 21. e, h, i; 22. c; 23. e; 24. b; 25. d; 26. a; 27. f; 28. organic; 29. shells; 30. bones; 31. atmospheric CO_2 gas; 32. photosynthesis; 33. aerobic respiration; 34. solar energy; 35. carbohydrates; 36. aerobic respiration; 37. carbohydrates; 38. CO_2; 39. atmosphere; 40. destroyed; 41. chloroplasts; 42. Calvin; 43. exchange, biomass; 44. bacteria; 45. plants; 46. nitrogen gas; 47. sedimentary; 48. because some usable energy is lost each time energy is transferred from one organism to another; 49. because the chemicals that compose living organisms are constantly cycled between organisms and the earth without loss of usable matter

Chapter Test

1. b; 2. b; 3. c; 4. b; 5. d; 6. a; 7. c; 8. b; 9. a; 10. a; 11. b; 12. d; 13. c; 14. a; 15. d; 16. a; 17. d; 18. d; 19. d; 20. c; 21. c; 22. c; 23. d; 24. b; 25. d

Chapter 24

Sections 24.1

1. d; 2. a; 3. f; 4. b; 5. g; 6. e; 7. c; 8. decrease, increase; 9. declining; 10. decrease, increase; 11. 3.5, 2.1; 12. less; 13. demography; 14. industrial, economic; 15. declining, higher; 16. equals; 17. population age structure; 18. less industrialized country; 19. there are more women entering reproductive age than leaving; 20. it increases the carrying capacity

Sections 24.2, 24.3, 24.4

1. e; 2. e; 3. i; 4. h; 5. j; 6. j; 7. a; 8. d; 9. b; 10. f; 11. c; 12. burning of vegetation, combustion; 13. burning vegetation releases CO_2 from the vegetation and reduces the number of plants available for CO_2 uptake in photosynthesis, combustion releases CO_2 from fossil fuels; 14. increase; 15. temperatures may increase; 16. stratosphere; 17. troposphere; 18. decrease; 19. increase; 20. sulfur dioxide; 21. smog formation; 22. less than; 23. increase, decreasing; 24. converts freshwater into; 25. cannot; 26. more; 27. do; 28. decreases

Sections 24.5, 24.6, 24.7

1. retrieving and transporting them has an environmental cost; 2. may contribute to greenhouse gases; 3. coal contains sulfur and may lead to formation of acid rain; 4. fossil fuels are a nonrenewable resource; 5. expensive; 6. potentially dangerous; 7. generates hazardous waste; 8. environmental

cost; **9.** contribute to pollution; **10.** not always available; **11.** solar power technology is not capable of meeting the energy demand; **12.** pollution, destruction of habitats, exploitation of resources, see sec. 24.6 for additional possible answers; **13.** consume less, recycle more, lower the worldwide fertility rate, reduce world poverty

Chapter Test

Multiple Choice
1. b; **2.** c; **3.** c; **4.** d; **5.** d; **6.** a; **7.** b; **8.** b; **9.** d; **10.** c; **11.** d; **12.** b; **13.** d; **14.** b; **15.** b; **16.** e; **17.** c; **18.** a; **19.** c; **20.** a; **21.** c; **22.** c; **23.** c; **24.** a; **25.** c